SpringerBriefs in Cell Biology

T0183923

For further volumes:
http://www.springer.com/series/10708

Jeffrey A. Stuart • Ellen L. Robb

Bioactive Polyphenols from Wine Grapes

 Springer

Jeffrey A. Stuart
Department of Biological Sciences
Brock University
St. Catharines, ON, Canada

Ellen L. Robb
Department of Biological Sciences
Brock University
St. Catharines, ON, Canada

ISBN 978-1-4614-6967-4   ISBN 978-1-4614-6968-1 (eBook)
DOI 10.1007/978-1-4614-6968-1
Springer New York Heidelberg Dordrecht London

Library of Congress Control Number: 2013934545

Printed on acid-free paper

Springer is part of Springer Science+Business Media (www.springer.com)

Preface

As biomedical researchers working and living in Niagara, Canada's preeminent wine region (apologies to the Okanagan Valley, Prince Edward County, and the Annapolis Valley), we have often been struck by how divergently researchers working in different fields approach the biology of grapevine polyphenols. Plant physiologists focus their efforts on methods for isolating and quantifying these molecules. Plant molecular biologists attempt to map the synthetic pathways responsible for the production of grapevine polyphenols, and transgenically augment their production, sometimes in other plant species, to assist in their further development into pharmaceuticals. Grape growers are interested in all of these activities, as they may lead to new markets and commercial applications for these molecules that are often found in abundance in some of the "waste" by-products of their annual harvest. Animal physiologists have revealed an increasing number of intriguing effects that might be harnessed to improve human health. Cell biologists have focused their attention on identifying specific molecular mechanisms that underlie the effects observed by their colleagues. At times, we have felt that researchers in these seemingly disparate fields may be tracking a parallel course, and that a convergence of these diverse perspectives would be advantageous to both the interpretation of experimental results and the process of experimental design. For these reasons, a more comprehensive overview seems certain to improve our understanding of these molecules, how they work, and how they might be fruitfully used by us. These thoughts were our motivation in undertaking this project.

Our goal here is to bring together the results of research in these various fields into a single resource to facilitate a more comprehensive understanding of the biological activities of the grapevine phytoalexins. We have focused on resveratrol and its derivatives since these have attracted by far the most attention amongst researchers for their ability to positively modulate human physiology. However, we hope that this resource will aid researchers in recognizing the many molecules beyond resveratrol with the potential to be studied and perhaps developed into nutraceuticals. To this end we feel that the book comes at an important moment in the field, as resveratrol derivatives, such as pterostilbene, piceid, and viniferins, seem to be on the cusp of attracting the kind of attention hitherto reserved for resveratrol itself.

Indeed, we hope that this book will serve as a catalyst for more research on the biological activities of these structurally related molecules in mammals, particularly humans, since in many instances they are both more abundant than resveratrol in grapevine tissues, are less rapidly degraded and excreted in vivo, and appear capable of eliciting very similar biological activities in mammalian cells and tissues. We hope that readers will find this book to be of use in their own research and development endeavors. Cheers.

ON, Canada   Jeffrey A. Stuart
                                                                            Ellen L. Robb

Acknowledgements

We are grateful to Max Merilovich, Melissa Ferguson, and Leah Shaver for their help in assembling the many reference papers for this book, including those used in the tables in Chap. 2. We also express our sincere appreciation to Lucas Maddalena and Casey Christoff for their thorough editing of the first draft, and to anonymous reviewers for their efforts in improving the manuscript. We appreciate the many conversations on these topics and enthusiastic support from our colleagues in the Cold Climate Oenology and Viticulture Institute and Biological Sciences Department at Brock University, particularly Debbie Inglis, Gary Pickering, and Vince DeLuca.

Contents

Chapter 1
Resveratrol and Its Derivatives as Phytoalexins

1.1 General Introduction

In this chapter we begin by providing an overview of the biochemistry of phyto-alexin synthesis in *Vitis vinifera* and the significance of these molecules in plant physiology. We then discuss the concentrations of these molecules in wines, the main dietary source of phytoalexins produced by *Vitis vinifera*. Our goal in this first chapter is to cultivate an appreciation for red wine polyphenols that extends beyond their effects in mammalian cells.

1.2 Grapevine Phytoalexins

Phytoalexins are secondary metabolites produced by plants in response to biotic and abiotic stressors. Across plant species, an enormous range of phytoalexins is pro-duced, and these generally have antifungal and antimicrobial activities (Ahuja et al. 2012). Phytoalexins, and related compounds termed phytoanticipins (constitutively produced secondary metabolites with similar properties), have been studied for their possible beneficial effects on human health. Amongst these molecules are biologi-cally active compounds with the potential to be further developed into functional food ingredients or dietary supplements (Boue et al. 2009). In the grapevine *Vitis vinifera*, the predominant phytoalexins produced in response to stress belong to a family of compounds called stilbenes (Fig. 1.1), which are synthesized from the amino acid phenylalanine.

One of the main stilbenes produced in grapevines is *trans*-resveratrol (*trans*-3,5,4'-trihydroxystilbene), which has been the subject of intensive investigation over the past two decades owing to its apparent health promoting properties. Although red wines are perhaps the best known dietary sources of resveratrol and its derivatives, these molecules are produced by a variety of phylogenetically diverse plant species, including peanut (*Arachis hypogaea*), Japanese knotweed (*Fallopia*

J.A. Stuart and E.L. Robb, *Bioactive Polyphenols from Wine Grapes*,
SpringerBriefs in Cell Biology, DOI 10.1007/978-1-4614-6968-1_1,
© The Author(s) 2013

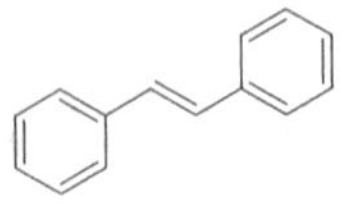

Fig. 1.1 Basic carbon skeleton structure of a stilbene

japonica), sorghum (*Sorghum bicolor*), and *Pinus* and *Picea* species (Parage et al. 2012). In instances where food or beverages are produced from these plants the biologically active phytoalexins are present in the consumed product, sometimes in negligible amounts and occasionally at concentrations sufficient to perhaps elicit biological responses.

1.3   Stilbene Synthesis

Trans-resveratrol is the initial stilbene product of *p*-coumaroyl-CoA and three molecules of malonyl-CoA in a reaction catalyzed by stilbene synthase (full pathway shown in Fig. 1.2). While most plants are capable of producing malonyl-CoA and CoA esters of cinnamic acid, the ability to synthesize stilbenes is limited to only a few plant species (reviewed in Chong et al. 2009). In *Vitis vinifera*, stilbene synthase is encoded by between 20 and 40 stilbene synthase genes, which is considerably more genetic diversity within this gene family than has been found in any other plant species (Chong et al. 2009; Parage et al. 2012). *Trans*-resveratrol serves as a precursor molecule that is further converted into a wide variety of compounds including piceid, pterostilbene, and viniferins by glycosylation, methoxylation, and oxidative oligomerization, respectively (Fig. 1.3). Some of these compounds, particularly piceid, are present at significantly higher concentrations than resveratrol in grapevine leaves (Boso et al. 2012), grape skins (Romero-Perez et al. 2001; Bavaresco et al. 2007), grape juices (Romero-Perez et al. 1999), and wines (Mark et al. 2005; Moreno-Labanda et al. 2004; Naugler et al. 2007).

Phytoalexins are produced in grapevines constitutively at low levels and are present in root, stem, leaf, and grape tissues (Wang et al. 2010). However, production and accumulation of phytoalexins within leaf and grape tissues is strongly induced by stressors such as UV light, heavy metals, ozone, and fungal infection (Jeandet et al. 2002). The expression of at least 20 stilbene synthase genes is induced by inoculation with the downy mildew fungus *Plasmopara viticola* (reviewed in Chong et al. 2009), and increased abundance of stilbene synthase protein on the leaf epidermis following UV exposure has been demonstrated using immunolocalization (Pan et al. 2009). *Trans*-resveratrol levels measured in leaf tissue vary with *Vitis vinifera* variety, but in all varieties they are increased substantially (up to 30-fold) within several days of inoculation with *P. viticola* (Boso et al. 2012). Like *trans-resveratrol*, the production of pterostilbene, piceid, and viniferins is also strongly enhanced by stress. Pterostilbene levels increase from undetectably low levels to a few µmol/mg

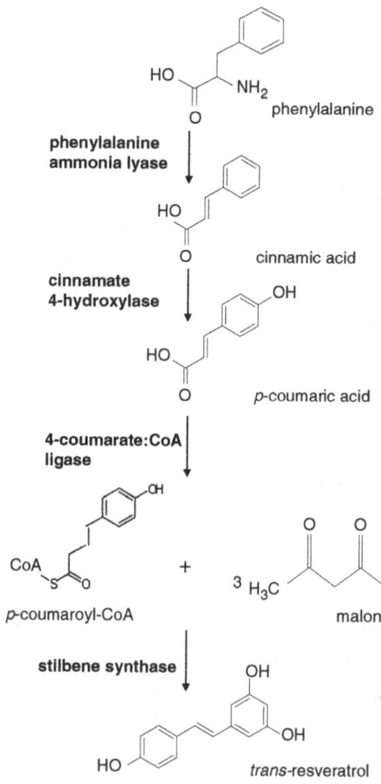

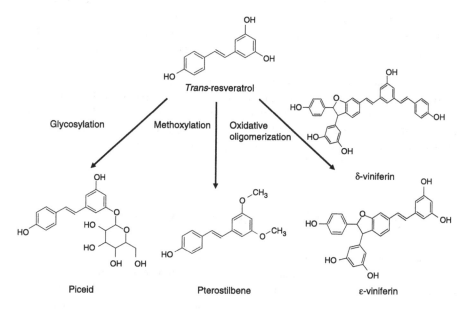

Fig. 1.2 Synthesis of *trans*-resveratrol from *p*-coumaroyl-CoA and malonyl-CoA by stilbene synthase

Fig. 1.3 *Trans*-resveratrol is a precursor molecule that may be converted into piceid, pterostilbene, and viniferins by glycosylation, methoxylation, and oxidative oligomerization, respectively

free weight within several days following infection with *P. viticola* (Boso et al. 2012), in parallel with an approximately fivefold increase in the expression of resveratrol O-methyltransferase, the enzyme catalyzing *trans*-pterostilbene synthesis from *trans*-resveratrol (Schmidlin et al. 2008). Piceid, ε-viniferin, and δ-viniferin undergo similar increases of up to 100-fold within 3 days following *P. viticola* infection. Although many of the measurements of polyphenol levels following infection have been made in grapevine leaves, similar increases have been shown also to occur in grape skins, which is relevant to the abundance of these molecules in red wines (Romero-Perez 2001; Montero et al. 2003).

1.4   Roles of Stilbenes in Grapevines

Stilbene production in plants is a central component of their response to stress. Resistance of *Vitis* species to fungal infection is generally correlated with their ability to produce stilbene phytoalexins (Douillet-Breuil et al. 1999; Schnee et al. 2008; Malacarne et al. 2011). Exogenous application of *trans*-resveratrol to grape berries is also an effective means of enhancing their resistance to fungal infection (Gonzalez Urena et al. 2003). Transgenic expression of stilbene synthase genes in non grapevine species is being utilized to enable stilbene production and in turn capture the potent antifungal properties of these molecules (Thomzik et al. 2001; Coutos-Thevenot et al. 2001; Zhu et al. 2004; Liu et al. 2011).

Indeed, there is good experimental evidence for potent antibacterial, antifungal, and anti-nematodal activities of *trans*-resveratrol and its derivatives (Chong et al. 2009). In laboratory experiments, *trans*-resveratrol inhibits *P. viticola* growth and development, and exhibits similar suppressive activity against a number of other fungal pathogens, including *Cladosporium cucumerinum, Botrytis cinerea, Oidium tuckeri, Pyricularia oryzae,* and *Sphaeropsis sapinea* at concentrations that are roughly equivalent to those produced by grapevine (reviewed in Jeandet et al. 2002). Pterostilbene is actually a more potent inhibitor of fungal growth than *trans-resveratrol* (Jeandet et al. 2002), though it may not accumulate to levels required for this activity in grape skin. ε-viniferin has antifungal activity similar to that of pterostilbene, suggesting that *trans*-resveratrol may be the least effective as an antifungal agent of this family of stilbene compounds.

1.5   Approaches to Increasing Stilbene Levels in Grapevines

Increased stilbene synthesis in *Vitis vinifera* is beneficial on two levels: (1) elevated stilbene levels, particularly in grape skin, will lead to higher concentrations of these apparent health-promoting compounds in wine, and (2) the increased stilbene concentrations will impart greater resistance to environmental stressors. Strategies to achieve this goal are therefore the focus of much research.

Application of exogenous stressors may be used to promote the endogenous production of *trans*-resveratrol and its derivatives in grapevine leaves and grapes, and this is of interest as a means of enhancing the levels of these compounds in red wines. One approach to enhancing stilbene levels in grapes has been to stimulate the signalling pathways involved in upregulating stilbene production. The plant hormone jasmonic acid is an effective inducer of phenylalanine ammonia lyase and stilbene synthase expression that significantly enhances stilbene synthesis. Repeated application of methyljasmonate to growing vines substantially increases the levels of *trans*-resveratrol and ε-viniferin in berries (Larronde et al. 2003; Vezzulli et al. 2007). In grapevine cell cultures, both methyljasmonate, and also cyclodextrins, stimulate stilbene synthesis (Lijavetzky et al. 2008; Zamboni et al. 2009). Interestingly, brief exposure to anoxia also appears to stimulate *trans*-resveratrol synthesis in harvested grapes (Jimenez et al. 2007). Thus, several non-genomic approaches have proven effective in enhancing stilbene levels in wine grapes and wine grape cells.

Another experimental method being used to increase stilbene levels is transgenic engineering to promote the synthesis of *trans*-resveratrol and its derivatives in plant species that do not naturally produce these molecules. Since all plant species produce the immediate precursors of *trans*-resveratrol, 4-coumaroyl-CoA and malonyl-CoA, transgenic overexpression of stilbene synthase alone is sufficient to instigate *trans*-resveratrol synthesis. Transgenic lines overexpressing stilbene synthase and/or O-methyltransferase have been engineered in a number of plant species, including tomato (*Solanum lycopersicum*), *Arabidopsis thaliana,* and tobacco (*Nicotiana tabacum*). Typically, when just stilbene synthase is overexpressed, the predominant polyphenol accumulating is *trans*-piceid rather than *trans*-resveratrol (Liu et al. 2006; Ingrosso et al. 2011). In transgenic plants co-expressing stilbene synthase and O-methyltransferase, pterostilbene accumulates as the major product, again with relatively low levels of *trans*-resveratrol (Rimando et al. 2012; Xu et al. 2012). Although these studies indicate that engineered plants could be used to produce stilbenes, it is interesting to note that there is evidence that excessive levels of these molecules interferes also with normal plant development (Ingrosso et al. 2011). The application of transgenic engineering to enhance stilbene levels is an active and developing area of research.

1.6   Stilbenes in Wines

Although resveratrol and its derivative molecules are found in a variety of wines, they are typically present at much higher levels in red wines due to a long fermentation process that includes contact with the grape skins. This allows the highly hydrophobic stilbene compounds to be extracted from grape skin into the forming ethanol. Although red wines are a relatively rich dietary source of resveratrol and its derivatives, the absolute concentrations of these compounds are nonetheless low, ranging from undetectable to about 63 μM (Stervbo et al. 2007; Dourtoglou et al. 1999;

Naugler et al. 2007). Levels of resveratrol vary with wine variety, with relatively high levels identified in some Pinot Noir and Merlot wines (Stervbo et al. 2007). Regional variation is also evident, and indeed environmental stresses unique to a given region are likely to affect resveratrol and stilbene production in grape skin, and therefore the levels present in wines produced from these grapes.

There has been a great deal of focus on the levels of resveratrol alone in red wines; however, it is important to note that some resveratrol derivatives are actually present at higher levels than resveratrol itself in red wines. For example, in studies of Hungarian (Mark et al. 2005), Spanish (Moreno-Labanda et al. 2004), and Canadian (Naugler et al. 2007) wines, piceid levels were found to be as much as tenfold higher than those of resveratrol. Piceid is also present at higher concentrations than resveratrol in grape juices (Romero-Perez et al. 1999) and cocoa (Hurst et al. 2008). Several resveratrol oligomers similarly accumulate to quite high levels in grapevine leaves (Boso et al. 2012), particularly following fungal elicitation. However, the levels of ε-viniferin (Adrian et al. 2000a, b), hopeaphenol (Boutegrabet et al. 2011), and pallidol (Naugler et al. 2007) in red wines appear to be generally lower than those of resveratrol and piceid. Pterostilbene levels are particularly low in grape berries and in red wines (e.g., Adrian et al. 2000a, b; Boso et al. 2012). Thus, resveratrol and piceid appear to be the major stilbenes present in red wines.

1.7 Conclusions

Grapevines produce a variety of stilbene molecules in response to both biotic and abiotic stresses. Identification of the signalling pathways and genes involved has led to strategies for targeting these with the goal of enhancing stilbene production both in *Vitaceae* and other plant species. Many of these strategies have now been shown to be quite effective at boosting the concentrations of resveratrol and its derivatives in wine grapes and instigating their production in other plant species. Given the accumulating evidence for applications of these compounds in human health contexts, ongoing research in this area is likely to yield increasingly high resveratrol grapes for wine production. Recent work has indicated that red wine supplemented with up to 200 mg/L resveratrol is palatable and stable (Gaudette & Pickering 2011), and therefore, high resveratrol wines should be an effective means of increasing dietary intake of resveratrol and its stilbene derivatives.

References

Adrian M, Jeandet P, Breuil AC, Levite D, Debord S, Bessis R (2000a) Assay of resveratrol and derivative stilbenes in wines by direct injection high performance liquid chromatography. Am J Enol Viticult 51:37–41
Adrian M, Jeandet P, Douillet-Breuil AC, Tesson L, Bessis R (2000b) Stilbene content of mature *Vitis vinifera* berries in response to UV-C elicitation. J Agric Food Chem 48:6103–6105

Ahuja I, Kissen R, Bones AM (2012) Phytoalexins in defense against pathogens. Trends Plant Sci 17:73–90

Bavaresco L, Pezzutto S, Gatti M, Mattivi F (2007) Role of the variety and some environmental factors on grape stilbenes. Vitis 46:57–61

Boso S, Alonso-Villaverde V, Martinez M-C, Kassemeyer H-H (2012) Quantification of stilbenes in *Vitis* genotypes with different levels of resistance to *Plasmopara viticola* infection. Am J Enol Viticult 63:419–423

Boue SM, Cleveland TE, Carter-Wientjes C, Shih BY, Bhatnagar D, McLachlan JM, Burow ME (2009) Phytoalexin-enriched functional foods. J Agric Food Chem 57:2614–2622

Boutegrabet L, Fekete A, Hertkorn N, Papastamoulis Y, Waffo-Téguo P, Mérillon JM, Jeandet P, Gougeon RD, Schmitt-Kopplin P (2011) Determination of stilbene derivatives in Burgundy red wines by ultra-high-pressure liquid chromatography. Anal Bioanal Chem 401:1513–1521

Chong J, Poutaraud A, Hugueney P (2009) Metabolism and roles of stilbenes in plants. Plant Sci 177:143–155

Coutos-Thevenot P, Poinssot B, Bonomelli A, Yean H, Breda C, Buffard D, Esnault R, Hain R, Boulay M (2001) In vitro tolerance to *Botrytis cinerea* of grapevine 41B rootstock in transgenic plants expressing the stilbene synthase *Vst*1 gene under the control of a pathogen-inducible PR10 promoter. J Exp Bot 52:901–910

Douillet-Breuil AC, Jeandet P, Adrian M, Bessis R (1999) Changes in the phytoalexin content of various Vits spp. In response to ultraviolet C elicitation. J Agric Food Chem 47:4456–4461

Dourtoglou VG, Makris DP, Bois-Dounasand F, Zonas C (1999) Trans-resveratrol concentration in wines produced in Greece. J Food Comps Anal 12:227–233

Gaudette NJ, Pickering GJ (2011) Sensory and chemical characteristics of trans-resveratrol fortified wine. Aust J Grape Wine Res 17:249–257

Gonzalez Urena A, Orea JM, Montero C, Jimenez JB, Gonzalez JL, Sanchez A, Dorado M (2003) Improving the post-harvest resistance in fruits by external application of trans-resveratrol. J Agric Food Chem 51:82–89

Hurst WJ, Glinski JA, Miller KB, Apgar J, Davey MH, Stuart DA (2008) Survey of the trans-resveratrol and trans-piceid content of cocoa-containing and chocolate products. J Agric Food Chem 56:8374–8378

Ingrosso I, Bonsegna S, De Demonico S, Laddomada B, Blando F, Santino A, Giovinazzo G (2011) Over-expression of a grape stilbene synthase gene in tomato induces parthenocarpy and causes abnormal pollen development. Plant Physiol Biochem 49:1092–1099

Jeandet P, Douillet-Breuil A-C, Bessis R, Debord S, Sbaghi M, Adrian M (2002) Phytoalexins from the Vitaceae: biosynthesis, phytoalexin gene expression in transgenic plants, antifungal activity, and metabolism. J Agric Food Chem 50:2731–2741

Jimenez JB, Orea JM, Urena AG, Escribano P, de la Osa PL, Guadarrama A (2007) Short anoxic treatments to enhance trans-resveratrol content in grapes and wine. Eur Food Res Technol 224:373–378

Larronde F, Gaudilliere JP, Krisa S, Decendit A, Deffieux G, Merillon JM (2003) Airborne methyl jasmonate induces stilbene accumulation in leaves and berries of grapevine plants. Am J Enol Viticult 54:63–66

Lijavetzky D, Almagro L, Belchi-Navarro S, Martinez-Zapater JM, Bru L, Pedreno MA (2008) Synergistic effect of methyljasmonate and cyclodextrin on stilbene biosynthesis pathway gene expression and resveratrol production in Monastrell grapevine cell cultures. BMC Res Notes 1:132

Liu SJ, Hu YL, Wang XL, Zhong J, Lin ZP (2006) High content of resveratrol in lettuce transformed with a stilbene synthase gene of Parthenocissus henryana. Journal of Agricultural and Food Chemistry 54:8082–8085

Liu Z, Zhuang C, Sheng S, Shao L, Zhao W, Zhao S (2011) Overexpression of a resveratrol synthase gene (*PcRS*) from *Polygonum cuspidatum* in transgenic *Arabidopsis* causes the accumulation of *trans*-piceid with antifungal activity. Plant Cell Rep 30:2027–2036

Malacarne G, Vrhovsek U, Zulini L, Cestaro A, Stefanini M, Mattivi F, Delledonne M, Velasco R, Moser C (2011) Resistance to Plasmopara viticola in a grapevine segregating population is associated with stilbenoid accumulation and with specific host transcriptional responses. BMC Plant Biol 11:114

Mark L, Nikfardjam MS, Avar P, Ohmacht R (2005) A validated HPLC method for the analysis of *trans*-resveratrol and *trans*-piceid in Hungarian wines. J Chromatogr Sci 43:445–449

Moreno-Labanda JF, Mallavia R, Pérez-Fons L, Lizama V, Saura D, Micol V (2004) Determination of piceid and resveratrol in Spanish wines deriving from Monastrell (*Vitis vinifera L.*) grape variety. J Agric Food Chem 52:5396–5403

Montero C, Cristescu SM, Jime'nez JB, Orea JM, te Lintel Hekkert S, Harren FJM, Gonza'lez Uren A (2003) Trans-resveratrol and grape disease resistance. A dynamical study by high-resolution laser-based techniques. Plant Physiol 131:129–138

Naugler C, McCallum JL, Klassen G, Strommer J (2007) Concentrations of *trans*-resveratrol and related stilbenes in Nova Scotia wines. Am J Enol Vit 58:117–119

Pan Q-H, Wang L, Li J-M (2009) Amounts and subcellular localization of stilbene synthase in response of grape berries to UV irradiation. Plant Sci 176:360–366

Parage C, Tavares R, Rety S, Baltenweck-Guyot R, Poutaraud A, Renault L, Heintz D, Lugan R, Marais G, Aubourg S, Hugueney P (2012) Structural, functional and evolutionary analysis of the unusually large stilbene synthase gene family in grapevine (*Vitis vinifera*). Plant Physiol 160(3):1407–1419

Rimando AM, Pan Z, Polashock JJ, Dayan FE, Mizuno CS, Snook ME, Liu C-J, Baserson SR (2012) *In planta* production of the highly potent resveratrol analogue pterostilbene via stilbene synthase and O-methyltransferase co-expression. Plant Biotech J 10:269–283

Romero-Pérez AI, Ibern-Gómez M, Lamuela-Raventós RM, de La Torre-Boronat MC (1999) Piceid, the major resveratrol derivative in grape juices. J Agric Food Chem 47:1533–1536

Romero-Pérez AI, Lamuela-Raventós RM, Andrés-Lacueva C, de La Torre-Boronat MC (2001) Method for the quantitative extraction of resveratrol and piceid isomers in grape berry skins. Effect of powdery mildew on the stilbene content. J Agric Food Chem 49:210–215

Schmidlin L, Poutaraud A, Claudel P, Mestre P, Prado E, Santos-Rosa M, Wiedemann-Merdinoglu S, Karst F, Merdinolglu D, Hugueney P (2008) A stress-inducible resveratrol O-methyltransferase involved in the biosynthesis of pterostilbene in grapevine. Plant Physiol 148:1630–1639

Schnee S, Viret O, Gindro K (2008) Role of stilbenes in the resistance of grapevine to powdery mildew. Physiol Mol Plant Pathol 72:128–133

Stervbo U, Vang O, Bonnesen C (2007) A review of the content of the putative chemopreventive phytoalexin resveratrol in red wine. Food Chem 101:449–457

Thomzik JE, Stenzel K, Stocker R, Schreier PH, Hain R, Stahl DJ (2001) Synthesis of a grapevine phytoalexin in transgenic tomatoes (*Lycopersicon Esculentum* Mill.) conditions resistance against *Phytophthora infestans*. Physiol Mol Plant Pathol 51:265–278

Vezzulli S, Civardi S, Ferrari F, Bavaresco L (2007) Methyl jasmonate treatment as a trigger of resveratrol synthesis in cultivated grapevine. Am J Enol Viticult 58:530–533

Wang W, Tang K, Yang HR, Wen PF, Zhang P, Wang HL, Huang WD (2010) Distribution of resveratrol and stilbene synthase in young grape plants (*Vitis vinifera* L. cv. Cabernet Sauvignon) and the effect of UV-C on its accumulation. Plant Physiol Biochem 48:142–152

Xu Y, Xu TF, Zhao XC, Zou Y, Li ZQ, Xiang J, Li FJ, Wang YJ (2012) Co-expression of VpROMT gene from Chinese wild Vitis pseudoreticulata with VpSTS in tobacco plants and its effects on the accumulation of pterostilbene. Protoplasma 249:819–833

Zamboni A, Gatto P, Cestaro A, Pilati S, Viola R, Mattivi F, Moser C, Velasco R (2009) Grapevine cell early activation of specific responses to DIMEB, a resveratrol elicitor. BMC Genomics 10:363

Zhu YJ, Agbayani R, Jackson MC, Tang CS, Moore PH (2004) Expression of the grapevine stilbene synthase gene VST1 in papaya provides increased resistance against diseases caused by *Phytophthora palmivora*. Planta 220:241–250

Chapter 2
Health Effects of Resveratrol and Its Derivatives

2.1 General Introduction

In this chapter we provide an overview of evidence for and against the putative beneficial effects of resveratrol and its derivatives in human health. Relevant data come from in vitro studies with human cells in culture, in vivo studies using primarily rodent models, and, where available, human clinical studies. While the detailed mechanisms underlying the observed effects are not discussed in detail here, these are the subject of considerable debate and are covered in Chap. 3.

2.2 Resveratrol and Cardiovascular Health

Resveratrol was initially identified as a bioactive component of red wines responsible for the "French Paradox," i.e., an observation that cardiovascular disease is less common in the French than predicted on the basis of dietary intake of saturated fats (Renaud and de Lorgeril 1992; reviewed in Soleas et al. 1997). Thus, research into the health effects of resveratrol as an isolated compound was initially focused on its interactions with vascular endothelium and platelets to affect aggregation and deposition reactions. The cardioprotective effects that have been ascribed to resveratrol include an ability to reduce the severity of damage incurred following a myocardial infarction, antiatherogenic effects, and positive effects on blood lipid profiles. Although the observed cardiovascular effects were initially attributed to resveratrol's chemical antioxidant capacity, more current research has focused on its ability to upregulate the expression of endogenous antioxidant enzymes, inhibit the inflammatory activity of cyclo-oxygenases, and promote nitric oxide signalling and vasodilation by activating nitric oxide synthases (reviewed in Ramprasath and Jones 2010). There is a paucity of data describing the cardiovascular effects of the other more abundant stilbenes present in red wines. However, it appears that piceid elicits a very similar range of cardiovascular effects, including protection against ischemia

J.A. Stuart and E.L. Robb, *Bioactive Polyphenols from Wine Grapes*,
SpringerBriefs in Cell Biology, DOI 10.1007/978-1-4614-6968-1_2,
© The Author(s) 2013

reperfusion events and vasodilation (reviewed in Long-tao et al. 2012). Few studies have focused on pterostilbene's activities in the cardiovascular system, but the limited data available does support the idea that it may also be cardioprotective (e.g., Park et al. 2010). Understanding the actions of the host of stilbenes found in red wine will be a valuable contribution to the data describing the cardiovascular effects of red wines.

The cardiovascular effects of resveratrol established primarily in rodent models have also been the focus of human clinical trials. In arteries isolated from the adipose tissue of normal human males, resveratrol induces relaxation and thus vasodilation at micromolar concentrations (Cruz et al. 2006). Similar effects have been shown with unpurified red wine polyphenol extract, which causes vasodilation of the brachial artery in males with coronary heart disease (Lekakis et al. 2005). Chronic dietary supplementation with a resveratrol-rich grape supplement attenuates a variety of pro-inflammatory markers in humans with cardiovascular disease (Tomé-Carneiro et al. 2012a, b). In humans, the levels of resveratrol metabolites in urine, indicating relatively high dietary intake, correlate with biomarkers of cardiovascular health (Zamora-Ros et al. 2012).

The molecular mechanisms underlying these cardiovascular effects appear to include interactions with cardiomyocytes and the modulation of nitric oxide signalling in vascular endothelial and smooth muscle cells via regulation of nitric oxide synthase isoforms. These and other mechanisms are discussed in detail in Chap. 3.

2.3 Red Wine Polyphenols and Cancer

In 1997 the first major study of resveratrol's anticancer effects was published by Jang and colleagues, and since that time there have been over a 1,000 publications on this topic. Roles for resveratrol in the inhibition of tumor initiation, promotion, and progression have been reported (e.g., Jang et al. 1997). In vitro, resveratrol effectively slows the growth of a large number of individual cancer cell lines (Table 2.1). In vivo, dietary resveratrol supplementation has been shown to slow the growth of transplanted tumors (Table 2.1). There is far less data on the effects of resveratrol derivatives on cell growth, but the available data suggests that piceid and pterostilbene have similar anticancer activities. For example, piceid inhibited lung tumor growth in mice at oral doses of 300 mg/kg twice daily (Kimura and Okuda 2000). Pterostilbene has recently received considerable attention in this context, and appears to be an effective inhibitor of cell proliferation at low micromolar concentrations (e.g., Lin et al. 2012; Wang et al. 2012a; reviewed by McCormack and McFadden 2012). In our experiments with C2C12 myoblasts, primary myoblasts, PC3 prostate cancer cells, SH-SY5Y neuroblastoma cells, and primary fibroblasts, both pterostilbene and piceid are at least as effective at inhibiting cell growth in vitro as resveratrol (unpublished data). Interestingly, the oligomeric resveratrol derivatives, including α-viniferin, ε-viniferin, pallidol, and trans-miyabenol all inhibit

Table 2.1 Resveratrol and wine polyphenols affect cell proliferation in vivo and in vitro (references from within past 5 years)

	Cell line	Cancer type	Polyphenol used	Observation	Reference
In Vitro	K562 and K562/IMA-2	Leukemia cell line	Resveratrol	Inhibition of cell growth, increased apoptosis	Can et al. (2012)
	KG-1a	Human promyeloblastic leukemia cells	Resveratrol	Inhibition of cell growth	Hu et al. (2012)
	A375	Melanoma	Pterostilbene	Inhibition of cell growth and increased apoptosis	Mena et al. (2012)
	A549	Lung cancer			
	HT29	Colon cancer			
	MCF7	Lung cancer			
	HT-29 and COLO 201	Human colon cancer cells	Resveratrol	Increased apoptosis	Miki et al. (2012)
	Capan-2, Panc-28, and HPDE	Human pancreatic cancer cells	Resveratrol	Inhibition of cell growth and increased apoptosis	Shamim et al. (2012)
	SGC7901	Human gastric adenocarcinoma	Resveratrol	Increased apoptosis	Wang et al. (2012b)
	HL-1 NB	Tumoral cardiac cell line	Resveratrol	Inhibition of growth, loss of cell adhesion, increased apoptosis	Baarine et al. (2011)
	U87 MG and U118 MG	Human glioblastoma cells	Resveratrol	Inhibition of cell proliferation, induction of cellular senescence	Gao et al. (2011)
	MCF7	Human breast cancer			
	H1299	Non-small-cell lung carcinoma			
	PC3	Human prostate cancer cells			
	H4-II-E	Rat hepatoma cells	Resveratrol	Inhibition of cell growth	Villa-Cuestra et al. (2011)
	PC-3 and C4-2B	Prostate cancer cells	Resveratrol	Cell growth inhibition induction of apoptosis	Brizuela et al. (2010)
	PANC-1, BxPC-3 and AsPC-1	Pancreatic cancer cells	Resveratrol	Inhibition of cell growth; cell cycle arrest and apoptosis	Cui et al. (2010)
	Mz-ChA-1, HuCC-T1, CCLP1, and SG231	Human cholangiocarcinoma cells	Resveratrol	Decreased cell proliferation, increased apoptosis	Frampton et al. (2010)
	MCF7	Breast cancer cell	Resveratrol	*Significant increase in MCF7 tumor cells growth rates	Gakh et al. (2010)

(continued)

Table 2.1 (continued)

Cell line	Cancer type	Polyphenol used	Observation	Reference
B16/DOX	Doxorubicin-resistant B16 melanoma cell subline	Resveratrol	G1 cell cycle arrest and induction of apoptosis	Gatouillat et al. (2010)
Caco-2	Colon cancer cells	Resveratrol	Inhibition of proliferation	Lea et al. (2010)
YUZAZ6 M14 A375	Melanoma cells	Resveratrol	Dose-dependant decrease in metabolic activity, decreased cell viability	Trapp et al. (2010)
MDA-MB-231	Breast cancer cells	Combination of Resveratrol, quercetin, catechin	Induction of apoptosis	Castillo-Pichardo et al. (2012)
CWR22Rv1	Prostate cancer cell line	Resveratrol	Inhibition of proliferation	Hsieh (2009)
HeLa	Cervical cancer cell line	Resveratrol	Inhibition of proliferation, induction of S phase cell cycle arrest	Kramer and Wesierska-Gadek (2009)
MB-CSC	Medulloblastoma cancer stem cells	Resveratrol	Inhibition of proliferation and tumorigenicity	Lu et al. (2009)
A549 CH27	Human lung cancer cell lines	Resveratrol	Inhibit cell population growth and induce cell injury	Weng et al. (2009)
MDA-MB-435	Breast cancer cell line	Combined resveratrol, quercetin and catechin	Inhibition of proliferation and cell cycle progression	Castillo-Pichardo et al. (2009)
HT29 Caco2 HepG2 HuTu80	Human colon adenocarcinoma cancer cell line Colon cancer cells Hepatocarcinoma cells Duodenum adenocarcinoma cells	Grape seed extract and red wine polyphenolic compounds	Inhibition of proliferation, increased apoptosis	Leifert and Abeywardena (2008)
HT29	Colorectal cancer cells	Pterostilbene and quercetin	Inhibition of cell growth	Priego et al. (2008)

In vivo			
Nude mice, mammary tumor	Combination of grape polyphenols	Decreased tumor growth and metastasis	Castillo-Pichardo and Dharmawardhane (2012)
Female nude mice injected with U87 MG or U118 MG cells	Resveratrol	Reduced tumorigenicity	Gao et al. (2011)
Male, Balb/c mice	Resveratrol	Decreased tumor incidence	George et al. (2011)
Male NMRI/Nu mice implanted with B16/DOX cells	Resveratrol	Decreased tumor growth	Brizuela et al. (2010)
Nude mice injected with Mz-ChA-1 cells	Resveratrol	Decreased tumor growth	Frampton et al. (2010)
PC-3 xenograft in nude mice	Resveratrol	Decreased tumor growth	Ganapathy et al. (2010)
Female B6D2F1 mice	Resveratrol	Decreased tumor growth	Gatouillat et al. (2010)
Athymic nude mice	Resveratrol	Decreased tumor growth	Oi et al. (2010)
BALB/c mice injected with C26 colon carcinoma cells	Red wine polyphenols (RWP)	Decreased tumor growth, reduced tumor vascularization, reduced number of lung metastases	Walter et al. (2010)
Rat hepatocarcinogenesis in Sprague–Dawley rats	Resveratrol	Reduced tumor growth	Bishayee and Dhir (2009)
Nude mice	Combination of resveratrol, quercetin, and catechin	Decreased primary mammary tumor growth	Castillo-Pichardo et al. (2009)
Simian Virus-40T-antigen (SV-40 Tag) targeted probasin promoter rat model	Genistein and resveratrol (alone and in combination)	Reduced prostate cancer cell proliferation	Harper et al. (2009)
Mice with normal and deficient TLR4 function with DMBA-induced skin carcinogenesis	Resveratrol	Fewer tumors, reduced tumor growth, inhibited angiogenesis	Yusuf et al. (2009)
Euthymic nude mice transplanted with MDA-MB231 cells	Merlot grape polyphenols	Arrest of tumor development	Hakimuddin et al. (2008)

growth at low micromolar concentrations in multiple cell lines including HepG2 liver cells (Colin et al. 2008), colon tumor cells (Marel et al. 2008), and B lymphocytic leukemia cells (Billard et al. 2002).

To date there is limited data available from clinical trials of resveratrol's anticancer effects in humans. However, resveratrol's role as a putative anticancer agent in humans is supported by the results of a recent study in which patients with colorectal cancer were given an oral resveratrol treatment for 8 days. Doses of 0.5 g/day and 1.0 g/day resveratrol significantly reduced cell proliferation in cancerous colon tissue (Patel et al. 2010). Further research involving larger patient cohorts is necessary before resveratrol can be applied to the prevention and treatment of human cancers.

Thus, many of the polyphenolic compounds identified in red wines, including resveratrol, pterostilbene, viniferins, and piceid, can inhibit the growth of cancerous and normal cells in vitro, and tumor grafts in vivo. Since there are no reports of toxicity in humans, there is potential for their use as anticancer agents. However, additional research is required in this area, particularly given the estrogenic properties of these molecules (discussed in further detail in Chap. 3). The structurally related phytoestrogen genistein appears to affect normal development of rodents when dietary supplementation occurs in the neonatal period (reviewed in Jefferson et al. 2007, 2012). A full characterization of resveratrol's physiological effects with an appreciation for its estrogen properties is necessary.

2.4   Red Wine Polyphenols and Neuroprotection

Tredici and colleagues (1999) hypothesized that resveratrol possessed neuroprotective properties in parallel with its better characterized cardioprotective effects. This property of resveratrol was demonstrated in a rat model of in vivo excitotoxic brain damage, where it conferred significant protection against systemic kainic acid injection (Virgili and Contestabile 2000). A similar protective effect of resveratrol against neuronal death in rat models of cerebral ischemia was subsequently shown (Huang et al. 2001; Sinha et al. 2002). There have now been many reports of resveratrol's neuroprotective activities in a variety of contexts (Table 2.2). Although resveratrol's neuroprotection was initially linked to its chemical antioxidant capacity, more recent reports have explored its biological activities, including the modulation of heme oxygenase (Zhuang et al. 2003), matrix metalloproteinase (Gagliano et al. 2005), nitric oxide synthase (Bi et al. 2005), and AMP kinase (Dasgupta and Milbrandt 2007) activities. More recently, the neuroprotective capacity of piceid has been evaluated, with similar outcomes. Acute piceid administration is protective in a similar rat model of brain ischemia/reperfusion injury as investigated with resveratrol (Cheng et al. 2006; Ji et al. 2012). Surprisingly, although the viniferin oligomers of resveratrol show neuroprotective properties in the same rat models of stroke (Kim et al. 2012), there is as yet no data published for pterostilbene.

Table 2.2 Neuroprotection associated with chronic administration of resveratrol and its derivatives

Model	Treatment	Stressor	Effect	Reference
PC12 cells	6–56 h of 5–25 µM RES	Oxygen–glucose deprivation	Antiapoptotic	Agrawal et al. (2011)
Rat hippocampal slices	75–500 µM RES	Oxygen–glucose deprivation	Reduced cell death	Raval et al. (2006)
Mice	5 mg/kg RES i.v.	Middle cerebral artery occlusion	Neuroprotective	Shin et al. (2012)
Rats	30 mg/kg RES for 7 days, i.p.	Common carotid and vertebral artery occlusion	Reduced death hippocampal CA1 neurons	Simão et al. (2012a)
Rats	30 mg/kg RES for 7 days, i.p.	Common carotid and vertebral artery occlusion	Reduced astroglial and microglial activation	Simão et al. (2012b)
Rats	10, 50, and 100 mg/kg RES i.p.	Bilateral carotid artery occlusion	Protected mitochondrial function of hippocampal CA1 neurons	Della-Morte et al. (2009)
Rats	10–100 µM RES by i.p. injection	Asphyxial cardiac arrest	Neuroprotective	Della-Morte et al. (2009)
Mice	50–100 mg/kg/day RES for 1–2 weeks	MPTP	Prevented loss of DA neurons	Blanchet et al. (2008)
Mice	10–40 mg/kg/day RES for 10 weeks	6-OHDA	Reduced neuronal damage	Jin et al. (2008)
HT22 cells	100 mM RES	Glutamate toxicity	Reduced incidence of cell death	Kukui et al. (2010)
SH-SY5Y cells	24–72 h of 1–50 µM RES	H_2O_2, paraquat or MMS	Reduced incidence of cell death	Robb and Stuart (2011)
Mice	10 mg/kg/day RES i.p. for 9 weeks	Maneb—and paraquat-induced parkinsonism	Reduced the neurodegenerative and paraquat accumulation	Srivastava et al. (2012)
Rat cortical neuron–glia cultures	2 µM RES for 12 h	Prion protein peptide PrP (106-126) toxicity	Prevented PrP (106-126)-induced neuronal cell death	Jeong et al. (2012)
Mouse Alzheimer's model	300 mg/kg/day RES in diet	N.A.	Reduced plaque pathology	Karuppagounder et al. (2009)
Mouse model of multiple sclerosis	100–250 mg/kg/day RES gavage	N.A.	Delay of autoimmune encephalomyelitis, retinal ganglion protection	Fonseca-Kelly et al. (2012)
SAMP8 mouse Alzheimer's model	120 mg/kg pterostilbene for 8 weeks	N.A.	Neuroprotective	Chang et al. (2012)
Striatal precursor cells expressing Mutant Htt	1 µM trans-ε-viniferin for 24–48 h	N.A.	Reduced ROS accumulation, prevented mitochondrial dysfunction	Fu et al. (2012)

Pterostilbene and resveratrol have been investigated for their ability to ameliorate the age-associated decline of cognitive function. In 19-month-old Fisher 344 rats (considered old for this strain) dietary pterostilbene administration for 12–13 weeks improved performance in the Morris water maze test, which is considered a test of working memory (Joseph et al. 2008). Resveratrol similarly preserved working memory in aged mice administered the pro-inflammation agent lipopolysaccharide (Abraham and Johnson 2009), and in aged rats (Zhao et al. 2012). Twenty weeks of dietary resveratrol supplementation also prevented the cognitive deficits caused by high fat feeding in mice, a model of the "cafeteria diet" in humans (Jeon et al. 2012). On the other hand, Park et al. (2012) report a negative effect of dietary resveratrol supplementation on spatial learning and memory in young mice. In the primate *Microcebus murinus* (grey mouse lemur), dietary supplementation with resveratrol for 18 months improved working and spatial memory (Dal-Pan et al. 2011a). The potential for red wine polyphenols to confer protection against acute neuronal insults or to ameliorate the symptoms of chronic neurodegenerative diseases has not been investigated in humans. However, the evidence gathered to date appears promising for the ability of resveratrol and pterostilbene to prevent age-associated cognitive impairments but more work, particularly in humans and with grapevine polyphenols other than resveratrol, is still needed.

2.5   Red Wine Polyphenols and Energy Homeostasis

One of the most publicized health claims for resveratrol in the popular media is its ability to impact body composition and to improve the negative metabolic consequences of high fat diets. In 2006, two high-profile reports provided evidence for a beneficial effect of dietary resveratrol supplementation in male mice fed a high fat diet (Baur et al. 2006; Lagouge et al. 2006). In male mice consuming a high fat diet, resveratrol supplementation at 22.4 mg/kg/day reduced body weight gains and decreased the incidence of spontaneous death over 60 weeks (Baur et al. 2006). Numerous markers of physiological well-being were evaluated and found to be improved with resveratrol supplementation in these mice. In high fat diet fed mice receiving 400 mg/kg/day resveratrol supplementation, the diet-induced body weight gain was reduced significantly, as was overall percentage body fat (Lagouge et al. 2006). In this latter study, body temperature and energy expenditure were increased by resveratrol supplementation, as was apparent mitochondrial abundance in skeletal muscle. Again, a wide variety of indicators of metabolic health were found to be positively altered by resveratrol supplementation in these obese male mice. Generally, resveratrol supplementation in male mice appears to confer protection against many of the negative physiological effects of high fat feeding, including adipogenesis and systemic markers of inflammation (e.g., Kim et al. 2011; Jeon et al. 2012).

Resveratrol supplementation has recently been evaluated in primates. Mouse lemurs (*Microcebus murinus*) given 200 mg resveratrol/kg/day for 4 weeks during their seasonal body mass gain period (in preparation for winter) showed reduced body mass gain, increased resting metabolic rates, and elevated body temperatures, though no differences in activity were observed (Dal-Pan et al. 2010). Interestingly, in a yearlong study these results, obtained during the winter (short day) season, were different from those found in summer (long day), where no effect on body mass was observed (Dal-Pan et al. 2011b). In this latter study, resveratrol supplementation increased 24 h energy expenditure and resting metabolic rate.

These results from primates were consistent with the notion that resveratrol supplementation could be effective in treating human obesity, and results from a very few human studies have now been reported. In obese (mean body mass index ~ 31.5) human males taking 150 mg/day resveratrol supplements for 30 days (e.g., Timmers et al. 2011) similar changes in muscle mitochondrial metabolic parameters, including apparent increases in mitochondrial abundance, were observed as had been reported in male mice fed a high fat diet (Lagouge et al. 2006; Baur et al. 2006). This result might suggest increased energy expenditure in human males consuming resveratrol supplements; however, no changes in body mass, percentage body fat, or 24 h energy expenditure were observed. In postmenopausal women with normal BMIs taking 75 mg/day resveratrol for 12 weeks (Yoshino et al. 2012), no effects on body mass, percentage body fat, 24 h energy expenditure, or other markers of overall health were observed (Yoshino et al. 2012). Taken together, the few human studies completed to date offer somewhat equivocal support for resveratrol supplementation, though longer-term studies in individuals of unhealthy weight are awaited. Also, at this time we are aware of no reports of the effects of other grapevine polyphenols on obesity and overweight.

2.6 Red Wine Polyphenols and Diabetes

One of the more prominent sequelae of overweight and obesity is type 2 diabetes mellitus, and the potential benefit of dietary resveratrol supplementation in normalizing glucose dyshomeostasis and reducing the side effects of diabetes has been studied (reviewed in Szkudelski and Szkudelska 2011). In rodent models with genetically or chemically induced diabetes, dietary resveratrol reduces many of the cardiovascular side effects of diabetes (Thirunavukkarasu et al. 2007; Silan 2008). There is evidence in the same experimental models that resveratrol can reduce hyperglycemia (Palsamy and Subramanian 2008; Penumathasa et al. 2008). Cellular studies suggest that this anti-hyperglycemic effect could be mediated by a stimulation of glucose transporter activities (e.g., GLUT4; Penumathsa et al. 2008). Other in vitro cellular studies indicate effects on the stability and insulin secretion rates of pancreatic beta cells (Palsamy and Subramanian 2010). In a recent clinical trial, patients with type 2 diabetes mellitus were given 150 mg/day resveratrol supplements (Bhatt et al. 2012). This regimen improved a variety of cardiovascular and blood parameters, including mean systolic blood pressure. These data support a beneficial effect of resveratrol on energy homeotasis in humans.

2.7   Resveratrol and Lifespan

In 2003 Howitz et al. reported the ability of resveratrol to extend lifespan in the baker's yeast *Saccharomyces cerevisiae*. These unicellular yeast have been widely used as an experimental model of aging and longevity (reviewed in Kaberlein et al. 2007), despite sharing essentially no physiology with mammals. Interestingly, resveratrol was reported to impact replicative aging (number of daughter cells per mother cell), but not chronological aging (length of time yeast survive in the non-dividing state) (Howitz et al. 2003). This observation motivated a series of studies in invertebrate and vertebrate metazoan species. Wood et al. (2004) reported that resveratrol significantly extended lifespan in two well-studied models of aging, *Drosophila melanogaster* and *Caenorhabditis elegans*. However, subsequent attempts to repeat these results yielded equivocal outcomes (Table 2.3). Bass et al. (2007) were unable to show an effect of resveratrol on longevity in *D. melanogaster*, despite using the same strain and dietary supplementation protocol. Wang et al. (2013) have recently provided data suggesting that, under some specific dietary regimens that differ from that reported by Wood et al. (2004), the lifespan of female *D. melanogaster* can be marginally affected. In this study, no effects were observed in males. In other species of flies, the effects of resveratrol appear to also be quite variable. In another species of fruit fly, *Anastrepha ludens*, dietary resveratrol supplementation had no effect on longevity in males and virtually no effects in females (Zou et al. 2009). In the honeybee, resveratrol increases average lifespan (Rascon et al. 2012). The original lifespan extension result reported for *C. elegans* (Wood et al. 2004) has proven more robust, though the magnitude of the effect reported in most experiments is generally quite small. Bass et al. (2007) showed a very subtle, but positive, effect of resveratrol on *C. elegans* longevity. Subsequently, modest lifespan extensions in *C. elegans* have been demonstrated by other researchers (Gruber et al. 2007; Greer and Brunet 2009; Zarse et al. 2010).

Evidence for effects of resveratrol on aging and longevity in vertebrate species is more limited. The first vertebrate model of aging and longevity in which an effect of resveratrol on lifespan was demonstrated was the short-lived annual fish species *Nothobranchius furzeri*. In this species, a highly significant increase in lifespan of up to 50 % was associated with dietary resveratrol delivery (Valenzano et al. 2006). Lifespan extension has also been reported in the related species *N. guentheri* (Genade and Lang 2013; Yu and Li 2012). Results for mammalian species have been available only relatively recently. In large, multicenter studies of genetically heterogeneous mice, dietary delivery of resveratrol failed to increase lifespan in males or females (Miller et al. 2011; Strong et al. 2013). A long-term study of dietary resveratrol supplementation has been initiated in the primate species *Microcebus murinus* (Dal-Pan et al. 2011b), and this study should provide the best data with which to judge whether there is any potential for resveratrol to affect human longevity.

Table 2.3 Reported effects of resveratrol on lifespan in various metazoan species

Species	Effect on lifespan	Reference
Drosophila melanogaster	Extension	Wood et al. (2004)
D. melanogaster	No effect	Bass et al. (2007)
D. melanogaster	No effect in males	Wang et al. (2013)
	Extension in females	
Anastrepha ludens	No effect	Zou et al. (2009)
Caenorhabditis elegans	Robust extension	Wood et al. (2004)
C. elegans	Marginal effect	Bass et al. (2007)
C. elegans	Extension	Gruber et al. (2007)
C. elegans	Extension	Greer and Brunet (2009)
C. elegans	Extension	Zarse et al. (2010)
Nothobranchius furzeri	Extension	Valenzano et al. (2006)
N. guentheri	Extension	Genade and Lang (2013)
N. guentheri	Extension	Yu and Li (2012)
Mus Musculus	No effect	Miller et al. (2011)
M. Musculus	No effect	Strong et al. (2013)
Microcebus murinus	Study not yet concluded	Dal-Pan et al. (2011b)

2.8   Conclusions: Red Wine Polyphenols and Their Putative Health Effects

In the two decades since the first putative human health effects of resveratrol were hypothesized and reported, a vast wealth of data has accumulated on the subject. Only relatively recently, this literature has expanded to include other grapevine polyphenols. Sufficient data is now available to support clinical trials of resveratrol and other polyphenols, and several of these have been completed or are ongoing. Further development of grapevine polyphenols for human health applications will require continued research into the underlying cellular and molecular mechanisms of these compounds, and details of their bioavailability in vivo. These are the subjects of Chaps. 3 and 4, respectively.

References

Abraham J, Johnson RW (2009) Consuming a diet supplemented with resveratrol reduced infection-related neuroinflammation and deficits in working memory in aged mice. Rejuvenation Res 12:445–453

Agrawal M, Kumar V, Kashyap MP, Khanna VK (2011) Ischemic insult induced apoptosis changes in PC12 cells: protection by trans resveratrol. Eur J Pharmacol 666:5–11

Baarine M, Thandapilly S, Louis X, Mazue F, Yu L, Delmas D, Netticadan T, Lizard G, Latruffe N (2011) Pro-apoptotic versus anti-apoptotic properties of dietary resveratrol on tumoral and normal cardiac cells. Genes Nutr 6:161–169

Bass TM, Weinkove D, Houthoofd K, Gems D, Partridge L (2007) Effects of resveratrol on lifespan in Drosopholia melanogaster and Caenorhabditis elegans. Mech Ageing Dev 128:546–552

Baur JA, Pearson KJ, Price NL, Jamieson HA, Lerin C, Kalra A, Prabhu VV, Allard JS, Lopez-Lluch G, Lewis K, Pistell PJ, Poosala S, Becker KG, Boss O, Gwinn D, Wang M, Ramaswamy S, Fishbein KW, Spencer RG, Lakatta EG, Le Couteur D, Shaw RJ, Navas P, Puigserver P, Ingram DK, de Cabo R, Sinclair DA (2006) Resveratrol improves health and survival of mice on a high-calorie diet. Nature 444:337–342

Bhatt JK, Thomas S, Nanjan MJ (2012) Resveratrol supplementation improves glycemic control in type 2 diabetes mellitus. Nutr Res 32:537–541

Bi XL, Yang JY, Dong YX, Wang JM, Cui YH, Ikeshima T, Zhao YQ, Wu CF (2005) Resveratrol inhibits nitric oxide and TNF-alpha production by lipopolysaccharide-activated microglia. Int Immunopharmacol 5:185–193

Billard C, Izard JC, Roman V, Kern C, Mathiot C, Mentz F, Kolb JP (2002) Comparative antiproliferative and apoptotic effects of resveratrol, epsilon-viniferin and vine-shots derived polyphenols (vineatrols) on chronic B lymphocytic leukemia cells and normal human lymphocytes. Leuk Lymphoma 43:1991–2002

Bishayee S, Dhir N (2009) Resveratrol-mediated chemoprevention of diethylnitrosamine-initiated hepatocarcinogenesis: inhibition of cell proliferation and induction of apoptosis. Chem Biol Interact 179:131–144

Blanchet J, Longpré F, Bureau G, Morissette M, Dipaolo T, Bronchi G, Martinoli MG (2008) Resveratrol, a red wine polyphenol, protects dopaminergic neurons in MPTP-treated mice. Prog Neuropsychopharmacol Biol Psychiatry 32:1243–1250

Brizuela L, Dayon A, Doumerc N, Ader I, Golzio M, Izard J, Hara Y, Malavaud B, Cuvillier O (2010) The sphingosine kinase-1 survival pathway is a molecular target for the tumor-suppressive tea and wine polyphenols in prostate cancer. FASEB J 24(10):3882–3894

Can G, Cakir Z, Kartal M, Gunduz U, Baran Y (2012) Apoptotic effects of resveratrol, a grape polyphenol, on imatinib-sensitive and resistant K562 chronic myeloid leukemia cells. Anticancer Res 32:2673–2678

Castillo-Pichardo L, Dharmawardhane S (2012) Grape polyphenols inhibit akt/mammalian target of rapamycin signaling and potentiate the effects of gefitinib in breast cancer. Nutr Cancer 64:1058–1069

Castillo-Pichardo L, Martínez-Montemayor M, Martínez J, Wall K, Cubano L, Dharmawardhane S (2009) Inhibition of mammary tumor growth and metastases to bone and liver by dietary grape polyphenols. Clin Exp Metastasis 26:505–516

Chang J, Rimando A, Pallas M, Camins A, Porquet D, Reeves J, Shukitt-Hale B, Smith MA, Joseph JA, Casadesus G (2012) Low-dose pterostilbene, but not resveratrol, is a potent neuromodulator in aging and Alzheimer's disease. Neurobiol Aging 33:2062–2071

Cheng Y, Zhang HT, Sun L, Guo S, Ouyang S, Zhang Y, Xu J (2006) Involvement of cell adhesion molecules in polydatin protection of brain tissues from ischemia-reperfusion injury. Brain Res 1110:193–200

Colin D, Lancon A, Delmas D, Lizard G, Abrossinow J, Kahn E, Jannin B, Latruffe N (2008) Antiproliferative activities of resveratrol and related compounds in human hepatocyte derived HepG2 cells are associated with biochemical cell disturbance revealed by fluorescence analyses. Biochimie 90:1674–1684

Cruz MN, Luksha L, Logman H, Poston L, Agewall S, Kublickiene K (2006) Acute responses to phytoestrogens in small arteries from men with coronary heart disease. Am J Physiol Heart Circ Physiol 290:H1969–H1975

Cui J, Sun R, Yu Y, Gou S, Zhao G, Wang C (2010) Antiproliferative effect of resveratrol in pancreatic cancer cells. Phytother Res 24:1637–1644

Dal-Pan A, Blanc S, Aujard F (2010) Resveratrol suppresses body mass gain in a seasonal non-human primate model of obesity. BMC Physiol 10:11

Dal-Pan A, Pifferi F, Marchal J, Picq JL, Aujard F, RESTRIKAL Consortium (2011a) Cognitive performances are selectively enhanced during chronic caloric restriction or resveratrolsupplementation in a primate. PLoS One 6:e16581

Dal-Pan A, Terrien J, Pifferi F, Botalla R, Hardy I, Marchal J, Zahariev A, Chery I, Zizzari P, Perret M, Picq JL, Epelbaum J, Blanc S, Aujard F (2011b) Caloric restriction or resveratrol supplementation and ageing in a non-human primate: first-year outcome of the RESTRIKAL study in Microcebus murinus. Age 33:15–31

Dasgupta B, Milbrandt J (2007) Resveratrol stimulates AMP kinase activity in neurons. Proc Natl Acad Sci USA 104:7217–7222

Della-Morte D, Dave KR, DeFazio RA, Bao YC, Raval AP, Perez-Pinzon MA (2009) Resveratrol pretreatment protects rat brain from cerebral ischemic damage via a sirtuin 1-uncoupling protein 2 pathway. Neuroscience 159:993–1002

Fonseca-Kelly Z, Nassrallah M, Uribe J, Khan RS, Dine K, Dutt M, Shindler KS (2012) Resveratrol neuroprotection in a chronic mouse model of multiple sclerosis. Front Neurol 3:84

Frampton G, Lazcano E, Li H, Mohamad A, DeMorrow S (2010) Resveratrol enhances the sensitivity of cholangiocarcinoma to chemotherapeutic agents. Lab Invest 90:1325–1338

Fu J, Jin J, Cichewicz RH, Hageman SA, Ellis TK, Xiang L, Peng Q, Jiang M, Arbez N, Hotaling K, Ross CA, Duan W (2012) Trans-(-)- ε-viniferin increases mitochondrial sirtuin 3 (SIRT3), activates AMP-activated protein kinase (AMPK), and protects cells in model of huntington disease. J Biol Chem 287:24460–24472

Gagliano N, Moscheni C, Torri C, Magnani I, Bertelli AA, Gioia M (2005) Effect of resveratrol on matrix metalloproteinase-2 (MMP-2) and secreted protein acidic and rich in cysteine (SPARC) on human cultured glioblastoma cells. Biomed Pharmacother 59:359–364

Gakh A, Anisimova N, Kiselevsky M, Sadovnikov S, Stankov I, Yudin M, Rufanov K, Krasavin M, Sosnov A (2010) Dihydro-resveratrol—a potent dietary polyphenol. Bioorg Med Chem Lett 20:6149–6151

Ganapathy S, Chen Q, Singh K, Shankar S, Sriivastava R (2010) Resveratrol enhances antitumor activity of TRAIL in prostate cancer xenografts through activation of FOXO transcription factor. PLoS One 5(12):e15627

Gao Z, Xu M, Barnett T, Xu W (2011) Resveratrol induces cellular senescence with attenuated mono-ubiquitination of histone H2B in glioma cells. Biochem Biophys Res Commun 407:271–276

Gatouillat G, Balasse E, Joseph-Pietras D, Morjani H, Madoulet C (2010) Resveratrol induces cell-cycle disruption and apoptosis in chemoresistant B16 melanoma. J Cell Biochem 110:893–902

Genade T, Lang DM (2013) Resveratrol extends lifespan and preserves glia but not neurons of the Nothobranchius guentheri optic tectum. Exp Gerontol 48(2):202–212

George J, Singh M, Srivastava A, Bhui K, Roy P, Chaturvedi P, Shukla Y (2011) Resveratrol and black tea polyphenol combination synergistically suppress mouse skin tumors growth by inhibition of activated MAPKs and p53. PLoS One 6:e23395

Greer EL, Brunet A (2009) Different dietary restriction regimens extend lifespan by both independent and overlapping genetic pathways in C. elegans. Aging Cell 8:113–127

Gruber J, Tang SY, Halliwell B (2007) Evidence for a trade-off between survival and fitness caused by resveratrol treatment of Caenorhabditis elegans. Ann N Y Acad Sci 1100:530–542

Hakimuddin F, Tiwari K, Paliyath G, Mechling K (2008) Grape and wine polyphenols downregulate the expression of signal transduction genes and inhibit the growth of estrogen receptor-negative MDA-MB231 tumors in nu/nu mouse xenografts. Nutr Res 28:702–713

Harper C, Cook L, Patel B, Wang J, Altoum I, Arabshahi A, Shiral T, Lamartiniere C (2009) Genistein and resveratrol, alone and in combination, suppress prostate cancer in SV-40 tag rats. Prostate 69:1668–1682

Howitz KT, Bitterman KJ, Cohen HY, Lamming DW, Lavu S, Wood JG, Zipkin RE, Chung P, Kisielewski A, Zhang LL, Scherer B, Sinclair DA (2003) Small molecule activators of sirtuins extend Saccharomyces cerevisiae lifespan. Nature 425(6954):191–196

Hsieh T (2009) Antiproliferative effects of resveratrol and the mediating role of resveratrol targeting protein NQO2 in androgen receptor-positive, hormone-non-responsive CWR22Rv1 cells. Anticancer Res 29:3011–3017

Hu L, Cao D, Li Y, He Y, Guo K (2012) Resveratrol sensitized leukemia stem cell-like KG-1a cells to cytokine-induced killer cells-mediated cytolysis through NKG2D ligands and TRAIL receptors. Cancer Biol Ther 13:516–526

Huang SS, Tsai MC, Chih CL, Hung LM, Tsai SK (2001) Resveratrol reduction of infarct size in Long-Evans rats subjected to focal cerebral ischemia. Life Sci 69:1057–1065

Jang M, Cai L, Udeani GO, Slowing KV, Thomas CF, Beecher CW, Fong HH, Farnsworth NR, Kinghorn AD, Mehta RG, Moon RC, Pezzuto JM (1997) Cancer chemopreventive activity of resveratrol, a natural product derived from grapes. Science 275:218–220

Jefferson WN, Padilla-Banks E, Newbold RR (2007) Disruption of the developing female reproductive system by phytoestrogens: genistein as an example. Mol Nutr Food Res 51:832–844

Jefferson WN, Patisaul HB, Williams CJ (2012) Reproductive consequences of developmental phytoestrogen exposure. Reproduction 143:247–260

Jeon BT, Jeong EA, Shin HJ, Lee Y, Lee DH, Kim HJ, Kang SS, Cho GJ, Choi WS, Roh GS (2012) Resveratrol attenuates obesity-associated peripheral and central inflammation and improves memory deficit in mice fed a high-fat diet. Diabetes 61:1444–1454

Jeong JK, Moon MH, Bae BC, Lee YJ, Seol JW, Kang HS, Kim JS, Kang SJ, Park SY (2012) Autophagy induced by resveratrol prevents human prion protein-mediated neurotoxicity. Neurosci Res 73:99–105

Ji H, Zhang X, Du Y, Liu H, Li S, Li L (2012) Polydatin modulates inflammation by decreasing NF-κB activation and oxidative stress by increasing Gli1, Ptch1, SOD1 expression and ameliorates blood-brain barrier permeability for its neuroprotective effect in pMCAO rat brain. Brain Res Bull 87:50–59

Jin F, Wu Q, Lu YF, Gong QH, Sh JS (2008) Neuroprotective effect of resveratrol on 6-OHDA-induced Parkinson's disease in rats. Eur J Pharmacol 600:7882

Joseph JA, Fisher DR, Cheng V, Rimando AM, Shukitt-Hale B (2008) Cellular and behavioural effects of stilbene resveratrol analogues: implications for reducing the deleterious effects of aging. J Agric Food Chem 56:10544–10561

Kaeberlein M, Burtner CR, Kennedy BK (2007) Recent developments in yeast aging. PLoS Genet 3(5):e84

Karuppagounder SS, Pinto JT, Xu H, Beal MF, Gibson GE (2009) Dietary supplementation with resveratrol reduced plaque pathology in a transgenic model Alzheimer's disease. Neurochem Int 54:111–118

Kim S, Jin Y, Choi Y, Park T (2011) Resveratrol exerts anti-obesity effects via mechanisms involving down-regulation of adipogenic and inflammatory processes in mice. Biochem Pharmacol 81:1343–1351

Kim JY, Jeong HY, Lee HK, Kim S, Hwang BY, Bae K, Seong YH (2012) Neuroprotection of the leaf and stem of Vitis amurensis and their active compounds against ischemic brain damage in rats and excitotoxicity in cultured neurons. Phytomedicine 19:150–159

Kimura Y, Okuda H (2000) Effects of naturally occurring stilbene glucosides from medicinal plants and wine, on tumour growth and lung metastasis in Lewis lung carcinoma-bearing mice. J Pharm Pharmacol 52:1287–1295

Kramer M, Wesierska-Gadek J (2009) Monitoring of long-term effects of resveratrol on cell cycle progression of human HeLa cells after administration of a single dose. Ann N Y Acad Sci 1171:257–263

Kukui M, Choi HJ, Zhu BT (2010) Mechanism for the protective effect of resveratrol against oxidative stress-induced neuronal death. Free Radic Biol Med 49:800–813

Lagouge M, Argmann C, Gerhart-Hines Z, Meziane H, Lerin C, Daussin F, Messadeq N, Milne J, Lambert P, Elliott P, Geny B, Laakso M, Puigserver P, Auwerx J (2006) Resveratrol improves mitochondrial function and protects against metabolic disease by activating SIRT1 and PGC-1alpha. Cell 127:1109–1122

Lea M, Ibeh C, Han L, Desbordes C (2010) Inhibition of growth and induction of differentiation markers by polyphenolic molecules and histone deacetylase inhibitors in colon cancer cells. Anticancer Res 30:311–318

Leifert W, Abeywardena M (2008) Grape seed and red wine polyphenol extracts inhibit cellular cholesterol uptake, cell proliferation, and 5-lipoxygenase activity. Nutr Res 28:842–850

Lekakis J, Rallidis LS, Andreadou I, Vamvakou G, Kazantzoglou G, Magiatis P, Skaltsounis AL, Kremastinos DT (2005) Polyphenolic compounds from red grapes acutely improve endothelial function in patients with coronary heart disease. Eur J Cardiovasc Prev Rehabil 12: 596–600

Lin VC, Tsai YC, Lin JN, Fan LL, Pan MH, Ho CT, Wu JY, Way TD (2012) Activation of AMPK by pterostilbene suppresses lipogenesis and cell-cycle progression in p53 positive and negative human prostate cancer cells. J Agric Food Chem 60:6399–6407

Long-tao L, Gang G, Min W, Zhang W (2012) The progress of the research on cardiovascular effects and acting mechanism of polydatin. Chin J Integr Med 18:714–719

Lu K, Chen Y, Tsia P, Tsia M, Lee Y, Chiang C, Kao C, Chiou S, Ku H, Lin C, Chen Y (2009) Evaluation of radiotherapy effect in resveratrol-treated medulloblastoma cancer stem-like cells. Childs Nerv Syst 25:543–550

Marel AK, Lizard G, Izard JC, Latruffe N, Delmas D (2008) Inhibitory effects of trans-resveratrol analogs molecules on the proliferation and the cell cycle progression of human colon tumoral cells. Mol Nutr Food Res 52:538–548

McCormack D, McFadden D (2012) Pterostilbene and cancer: current review. J Surg Res 173:e53–e61

Mena S, Rodriguez M, Ponsoda X, Estrela J, Jäättela M, Ortega A (2012) Pterostilbene-induced tumor cytotoxicity: a lysosomal membrane permeabilization-dependent mechanism. PLoS One 7:e44524

Miki H, Uehara N, Kimura A, Sasaki T, Yuri T, Yoshizawa K, Tsubura A (2012) Resveratrol induces apoptosis via ROS-triggered autophagy in human colon cancer cells. Int J Oncol 40:1020–1028

Miller RA, Harrison DE, Astle CM, Baur JA, Boyd AR, de Cabo R, Fernandez E, Flurkey K, Javors MA, Nelson JF, Orihuela CJ, Pletcher S, Sharp ZD, Sinclair D, Starnes JW, Wilkinson JE, Nadon NL, Strong R (2011) Rapamycin, but not resveratrol or simvastatin, extends life span of genetically heterogeneous mice. J Gerontol A Biol Sci Med Sci 66:191–201

Oi N, Jeong C, Nadas J, Cho Y, Pugliese A, Bode A, Dong Z (2010) Resveratrol, a red wine polyphenol, suppresses pancreatic cancer by inhibiting leukotriene A_4 hydrolase. Cancer Res 70:9755–9764

Palsamy P, Subramanian S (2008) Resveratrol, a natural phytoalexin, normalizes hyperglycemia in streptozotocin-nicotinamide induced experimental diabetic rats. Biomed Pharmacother 62:598–605

Palsamy P, Subramanian S (2010) Ameliorative potential of resveratrol on proinflammatory cytokines, hyperglycemia mediated oxidative stress, and pancreatic beta-cell dysfunction in streptozotocin-nicotinamide-induced diabetic rats. J Cell Physiol 224:423–432

Park ES, Lim Y, Hong JT, Yoo HS, Lee CK, Pyo MY, Yun YP (2010) Pterostilbene, a natural dimethylated analog of resveratrol, inhibits rat aortic vascular smooth muscle cell proliferation by blocking Akt-dependent pathway. Vascul Pharmacol 53:61–67

Park HR, Kong KH, Yu BP, Mattson MP, Lee J (2012) Resveratrol inhibits the proliferation of neural progenitor cells and hippocampal neurogenesis. J Biol Chem 287:42588–42600

Patel KR, Brown VA, Jones DJ, Britton RG, Hemingway D, Miller AS, West KP, Booth TD, Perloff M, Crowell JA, Brenner DE, Steward WP, Gescher AJ, Brown K (2010) Clinical pharmacology of resveratrol and its metabolites in colorectal cancer patients. Cancer Res 70(19): 7392–7399

Penumathsa SV, Thirunavukkarasu M, Zhan L, Maulik G, Menon VP, Bagchi D, Maulik N (2008) Resveratrol enhances GLUT-4 translocation to the caveolar lipid raft fractions through AMPK/Akt/eNOS signalling pathway in diabetic myocardium. J Cell Mol Med 12:2350–2361

Priego S, Feddi F, Ferrer P, Mena S, Beniloch M, Ortega A, Carretero J, Obrador E, Asensi M, Astrela J (2008) Natural polyphenols facilitate elimination of HT-29 colorectal cancer xenografts by chemoradiotherapy: a Bcl-2- and superoxide dismutase 2-dependent mechanism. Mol Cancer Ther 7:3330–3342

Ramprasath VR, Jones PJ (2010) Anti-atherogenic effects of resveratrol. Eur J Clin Nutr 64: 660–668

Rascón B, Hubbard BP, Sinclair DA, Amdam GV (2012) The lifespan extension effects of resveratrol are conserved in the honey bee and may be driven by a mechanism related to caloric restriction. Aging 4:499–508

Raval AP, Dave KR, Pérez-Pinzón MA (2006) Resveratrol mimics ischemic preconditioning in the brain. J Cereb Blood Flow Metab 26:1141–1147

Renaud S, de Lorgeril M (1992) Wine, alcohol, platelets, and the French paradox for coronary heart disease. Lancet 339:1523–1526

Robb EL, Stuart JA (2011) Resveratrol interacts with estrogen receptor-β to inhibit cell replicative growth and enhance stress resistance by upregulating mitochondrial superoxide dismutase. Free Radic Biol Med 50(7):821–831

Shamim U, Hanif S, Albanyan A, Beck F, Bao B, Wang Z, Banerjee S, Sarkar F, Mohammad R, Hadi S, Azmi A (2012) Resveratrol-induced apoptosis is enhanced in low pH environments associated with cancer. J Cell Physiol 227:1493–1500

Shin JA, Lee KE, Kim HS, Park EM (2012) Acute resveratrol treatment modulates multiple signaling pathways in the ischemic brain. Neurochem Res 37:2686–2696

Silan C (2008) The effects of chronic resveratrol treatment on vascular responsiveness of streptozotocin-induced diabetic rats. Biol Pharm Bull 31:897–902

Simão F, Matté A, Pagnussat AS, Netto CA, Salbego CG (2012a) Resveratrol prevents CA1 neurons against ischemic injury by parallel modulation of both GSK-3β and CREB through PI3-K/Akt pathways. Eur J Neurosci 36:2899–2905

Simão F, Matté A, Pagnussat AS, Netto CA, Salbego CG (2012b) Resveratrol preconditioning modulates inflammatory response in the rat hippocampus following global cerebral ischemia. Neurochem Int 61:659–665

Sinha K, Chaudhary G, Gupta YK (2002) Protective effect of resveratrol against oxidative stress in middle cerebral artery occlusion model of stroke in rats. Life Sci 71:655–665

Soleas GJ, Diamandis EP, Goldberg DM (1997) Resveratrol: a molecule whose time has come? And gone? Clin Biochem 30:91–113

Srivastava G, Dixit A, Yadav S, Patel DK, Prakash O, Singh MP (2012) Resveratrol potentiates cytochrome P450 2 d22-mediated neuroprotection in maneb- and paraquat-induced parkinsonism in the mouse. Free Radic Biol Med 52:1294–1306

Strong R, Miller RA, Astle CM, Baur JA, de Cabo R, Fernandez E, Guo W, Javors M, Kirkland JL, Nelson JF, Sinclair DA, Teter B, Williams D, Zaveri N, Nadon NL, Harrison DE (2013) Evaluation of resveratrol, green tea extract, curcumin, oxaloacetic acid, and medium-chain triglyceride oil on life span of genetically heterogeneous mice. J Gerontol A Biol Sci Med Sci 68(1):6–16

Szkudelski T, Szkudelska K (2011) Anti-diabetic effects of resveratrol. Ann N Y Acad Sci 1215:34–39

Thirunavukkarasu M, Penumathsa SV, Koneru S, Juhasz B, Zhan L, Otani H, Bagchi D, Das DK, Maulik N (2007) Resveratrol alleviates cardiac dysfunction in streptozotocin-induced diabetes: role of nitric oxide, thioredoxin, and heme oxygenase. Free Radic Biol Med 43:720–729

Timmers S, Konings E, Bilet L, Houtkooper RH, van de Weijer T, Goossens GH, Hoeks J, van der Krieken S, Ryu D, Kersten S, Moonen-Kornips E, Hesselink MK, Kunz I, Schrauwen-Hinderling VB, Blaak EE, Auwerx J, Schrauwen P (2011) Calorie restriction-like effects of 30 days of resveratrol supplementation on energy metabolism and metabolic profile in obese humans. Cell Metab 14:612–622

Tomé-Carneiro J, Gonzálvez M, Larrosa M, García-Almagro FJ, Avilés-Plaza F, Parra S, Yáñez-Gascón MJ, Ruiz-Ros JA, García-Conesa MT, Tomás-Barberán FA, Espín JC (2012a) Consumption of a grape extract supplement containing resveratrol decreases oxidized LDL and ApoB in patients undergoing primary prevention of cardiovascular disease: a triple-blind, 6-month follow-up, placebo-controlled, randomized trial. Mol Nutr Food Res 56:810–821

Tomé-Carneiro J, Gonzálvez M, Larrosa M, Yáñez-Gascón MJ, García-Almagro FJ, Ruiz-Ros JA, García-Conesa MT, Tomás-Barberán FA, Espín JC (2012b) One-year consumption of a grape nutraceutical containing resveratrol improves the inflammatory and fibrinolytic status of patients in primary prevention of cardiovascular disease. Am J Cardiol 110:356–363

Trapp V, Parmakhtiar B, Papazian V, Willmott L, Fruehauf J (2010) Anti-angiogenic effects of resveratrol mediated by decreased VEGF and increased TSP1 expression in melanoma-endothelial cell co-culture. Angiogenesis 13:305–315

Tredici G, Miloso M, Nicolini G, Galbiati S, Cavaletti G, Bertelli A (1999) Resveratrol, map kinases and neuronal cells: might wine be a neuroprotectant? Drugs Exp Clin Res 25:99–103

Valenzano DR, Terzibasi E, Genade T, Cattaneo A, Domenici L, Cellerino A (2006) Resveratrol prolongs lifespan and retards the onset of age-related markers in a short-lived vertebrate. Curr Biol 16:296–300

Villa-Cuestra E, Boylan J, Tatar M, Gruppuso P (2011) Resveratrol inhibits protein translation in hepatic cells. PLoS One 6:e29513

Virgili M, Contestabile A (2000) Partial neuroprotection of in vivo excitotoxic brain damage by chronic administration of the red wine antioxidant agent, trans-resveratrol in rats. Neurosci Lett 281:123–126

Walter A, Etienne-Selloum N, Brasse D, Khallouf H, Bronner C, Rio M, Beretz A, Schini-Kerth V (2010) Intake of grape-derived polyphenols reduces C26 tumor growth by inhibiting angiogenesis and inducing apoptosis. FASEB J 24:3360–3369

Wang Y, Ding L, Wang X, Zhang J, Han W, Feng L, Sun J, Jin H, Wang XJ (2012a) Pterostilbene simultaneously induces apoptosis, cell cycle arrest and cyto-protective autophagy in breast cancer cells. Am J Transl Res 4:44–51

Wang Z, Li W, Meng X, Jia B (2012b) Resveratrol induces gastric cancer cell apoptosis via reactive oxygen species, but independent of sirtuin1. Clin Exp Pharmacol Physiol 39:227–232

Wang C, Wheeler CT, Alberico T, Sun X, Seeberger J, Laslo M, Spangler E, Kern B, de Cabo R, Zou S (2013) The effect of resveratrol on lifespan depends on both gender and dietary nutrient composition in Drosophila melanogaster. Age (Dordr) 35(1):69–81

Weng C, Yang Y, Ho C, Yen G (2009) Mechanisms of apoptotic effects induced by resveratrol, dibenzoylmethane, and their analogues on human lung carcinoma cells. J Agric Food Chem 57:5235–5243

Wood JG, Rogina B, Lavu S, Howitz K, Helfand SL, Tatar M, Sinclair D (2004) Sirtuin activators mimic caloric restriction and delay ageing in metazoans. Nature 430:686–689

Yoshino J, Conte C, Fontana L, Mittendorfer B, Imai S, Schechtman KB, Gu C, Kunz I, Rossi Fanelli F, Patterson BW, Klein S (2012) Resveratrol supplementation does not improve metabolic function in nonobese women with normal glucose tolerance. Cell Metab 16:658–664

Yu X, Li G (2012) Effects of resveratrol on longevity, cognitive ability and aging-related histological markers in the annual fish Nothobranchius guentheri. Exp Gerontol 47:940–949

Yusuf N, Nasti T, Meleth S, Elmets C (2009) Resveratrol enhances cell-mediated immune response to DMBA through TLR4 and prevents DMBA induced cutaneous carcinogenesis. Mol Carcinog 48:713–723

Zamora-Ros R, Urpi-Sarda M, Lamuela-Raventós RM, Martínez-González MÁ, Salas-Salvadó J, Arós F, Fitó M, Lapetra J, Estruch R, Andres-Lacueva C, PREDIMED Study Investigators (2012) High urinary levels of resveratrol metabolites are associated with a reduction in the prevalence of cardiovascular risk factors in high-risk patients. Pharmacol Res 65:615–620

Zarse K, Schmeisser S, Birringer M, Falk E, Schmoll D, Ristow M (2010) Differential effects of resveratrol and SRT1720 on lifespan of adult Caenorhabditis elegans. Horm Metab Res 42:837–839

Zhao H, Niu Q, Li X, Liu T, Xu Y, Han H, Wang W, Fan N, Tian Q, Zhang H, Wang Z (2012) Long-term resveratrol consumption protects ovariectomized rats chronically treated with D-galactose from developing memory decline without effects on the uterus. Brain Res 1467:67–80

Zhuang H, Kim YS, Koehler RC, Doré S (2003) Potential mechanism by which resveratrol, a red wine constituent, protects neurons. Ann N Y Acad Sci 993:276–286

Zou S, Carey JR, Liedo P, Ingram DK, Müller HG, Wang JL, Yao F, Yu B, Zhou A (2009) The prolongevity effect of resveratrol depends on dietary composition and calorie intake in a tephritid fruit fly. Exp Gerontol 44:472–476

Chapter 3
Cellular and Molecular Mechanisms of Resveratrol and Its Derivatives

3.1 Introduction

The list of molecular targets ascribed to resveratrol (there are relatively limited data at this time for resveratrol derivatives) has grown considerably over the past decade, due to the great interest in this compound's putative health promoting effects. Direct physical interactions of resveratrol with estrogen receptors (ERs) (e.g., Gehm et al. 1997), protein deacetylases (e.g., SIRT1; reviewed in Hu et al. 2011), protein kinases (e.g., AMP kinase; Dasgupta and Milbrandt 2007), phosphodiesterase (e.g., Park et al. 2012), heat shock proteins (e.g., HSP25; Han et al. 2012), and regulators of cellular bioenergetics (e.g., PGC-1alpha; Lagouge et al. 2006) have all been reported. Resveratrol also exerts effects via transcriptional regulation of a wide range of genes, including nitric oxide synthases (e.g., Csiszar et al. 2009), p53 and other cell cycle regulatory proteins (e.g., Whyte et al. 2007), and the mitochondrial antioxidant enzyme manganese superoxide dismutase (MnSOD; e.g., Robb et al. 2008a, b; Robb and Stuart 2011). Given this preponderance of reported molecular targets it can be difficult to arrive at a satisfying understanding of the compound's biological activities in mammalian systems.

In our view, the most parsimonious explanation for resveratrol's many and wide-ranging effects is that it stimulates ER signalling pathways. ER signalling elicits both acute and longer-term effects and targets hundreds of individual genes and cellular processes (reviewed in Barros and Gustafsson 2011; Leitman et al. 2010). While this does not exclude activities via other mechanisms, there is nonetheless an overwhelming concordance between the cellular and systemic effects elicited by resveratrol and its derivatives, and those elicited by natural endogenous estrogens such as 17β-estradiol. The earliest reports of resveratrol's biological activities in animals showed it acting as an ER agonist (Gehm et al. 1997). Subsequently, resveratrol's ability to bind the classical ERs ERalpha and ERbeta has been demonstrated (Bowers et al. 2000) in vitro and in silico (Yuan et al. 2011). Some of resveratrol's important cellular effects are mediated by ERbeta (e.g., Robb and Stuart 2011, Robb and Stuart, unpublished data). In this chapter, we review the many of the reported

J.A. Stuart and E.L. Robb, *Bioactive Polyphenols from Wine Grapes*, SpringerBriefs in Cell Biology, DOI 10.1007/978-1-4614-6968-1_3, © The Author(s) 2013

molecular mechanisms underlying the biological activities of resveratrol and other grapevine polyphenols in mammalian cells, and compare them to those associated with estrogens.

3.2 Estrogen Receptors

The classical ERs, ERalpha and ERbeta, transcriptionally regulate hundreds of genes (reviewed in Leitman et al. 2010). ERalpha is highly expressed in reproductive tissues and ERalpha agonism plays a prominent role in reproductive physiology. On the other hand, ERbeta is expressed in many tissues of both males and females, including brain, heart, lung, epithelium, gastrointestinal tract, and prostate gland (see Nilsson et al. 2011 for review). Whereas ERalpha agonism is generally pro-proliferative, ERbeta agonists are typically anti-proliferative (Fig. 3.1) (Sugiyama et al. 2010). In addition to transcriptional effects, estrogens also exert acute effects via a G-protein coupled ER (Nilsson et al. 2011). Here we focus less on these acute short-term effects than on the longer-term transcriptional effects exerted by estrogens and red wine polyphenols, since a major use of these latter compounds is as dietary supplements, which would presumably be consumed chronically. For most of the putative health promoting activities discussed in Chap. 2, we draw comparisons between the effects elicited by resveratrol and those associated with estrogens.

3.3 Neuroprotective Mechanisms of Resveratrol, and the Parallels to Estrogen Signalling

There are dozens of papers describing the neuroprotective actions of both resveratrol and 17β-estradiol in essentially the same experimental models of brain injury and neurodegeneration. Treatment of rats with resveratrol by intraperitoneal injection for 7 days confers substantial protection in ischemic stroke models using

ERalpha	ERbeta
Breast, ovary, uterus*, testis, pituitary, brain, aorta, heart, skeletal muscle, kidney, pancreas, colon, small intestine, bone, skin * Highly abundant	Ovary,* uterus, testis*, pituitary, brain*, heart, skeletal muscle, kidney*, pancreas, colon, small intestine, bone*, skin* * Highly abundant
↑ Cell proliferation	↓ Cell proliferation

Fig. 3.1 Tissue expression of ERalpha compared to ERbeta (based on data from Brandenberger et al. 1997)

30 min of middle cerebral artery occlusion (Ren et al. 2011) or bilateral occlusion of the common carotid arteries for 10 min (Simão et al. 2011). Resveratrol pretreatment also protects against neuronal death and brain damage caused by brain ischemia secondary to cardiac arrest (Della-Morte et al. 2009). Similar neuroprotective effects of resveratrol have been shown in vitro. Pretreatment of PC12 neuronal cells with 25 μM resveratrol for 24 h prior to oxygen–glucose deprivation ameliorated oxidative damaged caused by 6 h oxygen–glucose deprivation (Agrawal et al. 2011). Similarly, pretreatment of rat hippocampal slices with 75–500 μM resveratrol prior to oxygen–glucose deprivation was protective against neuronal death (Raval et al. 2006).

Although far less data is available for resveratrol derivatives, recent evidence suggests similar neuroprotective ability. Neuroprotective effects of piceid were demonstrated in rats using a permanent middle cerebral artery occlusion (pMCAO) model of stroke that was coincident with piceid administration by intraperitoneal injection. Infarct volume and neurological deficit score were evaluated at 24 h and 72 h post pMCAO, and a dose of 50 mg/kg piceid was found to be protective (Ji et al. 2012). These results are generally in agreement with an earlier study (Cheng et al. 2006) showing a neuroprotective effect of 30 mg/kg piceid in a transient focal ischemia model. The effects of longer-term dosing with pterostilbene and viniferins appear not to have been tested in the context of acute ischemic brain injury.

These neuroprotective effects of resveratrol and its derivatives are strikingly similar to those reported for 17β-estradiol and specific ER subtype agonists. 17β-estradiol-mediated neuroprotection has been demonstrated in a wide variety of in vitro and in vivo experimental contexts, and protection from both acute trauma (e.g., ischemic/reperfusion injury; reviewed in Simpkins et al. 2012) and chronic degenerative diseases (Parkinson's disease; reviewed in Bourque et al. 2009) have been shown. In vitro models of ischemic trauma have included glutamate excitotoxicity and NMDA excitotoxicity. In cultured hippocampal slices, 24 h pretreatment with estrogen confers significant protection against NMDA toxicity (Aguirre and Baudry 2009). This result can be duplicated using the ERbeta-specific agonist diarylpropionitrile (DPN) but not with the ERalpha-specific agonist 4,4',4"-(4-Propyl-[1H]-pyrazole-1,3,5-triyl)trisphenol (PPT), suggesting that it is mediated by ERbeta. Subsequent work by this research group confirmed this in the same model system, showing the absence of effect in hippocampal slices from ERbeta knockout mice (Aguirre et al. 2010). In similar studies of hippocampal neuron cultures, a 48 h pretreatment with 17β-estradiol, DPN or PPT was shown to confer protection against glutamate excitotoxicity. Together these results indicate that pretreatment with ERbeta and ERalpha agonists can confer protection against neuronal death, though ERbeta may play a dominant role.

The effects of estrogens and selective estrogen receptor modulators (SERMs) in the brain are not specific to neurons. Pretreatment with 17β-estradiol protects glial cells against oxygen glucose deprivation-mediated cell death, though in this case ERalpha appears to play the dominant role since this protective effect is reproduced by the ERalpha-specific agonist PPT (Guo et al. 2010). Similarly, pretreatment of primary astrocyte cultures with 17β-estradiol for up to 48 h prevents cell death caused by OGD (Guo et al. 2012). Again, this effect can be duplicated using PPT but not with DPN, suggesting that ERalpha is the critical receptor involved.

Similar protective effects of estrogen and SERM pretreatment have been shown in vivo. Using subcutaneous implants, Horsburgh et al. (2002) delivered estrogen or vehicle to ovariectomized mice continuously for 2 weeks, then subjected them to 17 min of global cerebral ischemia, and allowed 72 h recovery. The authors found significantly less damage in brains of estrogen treated animals. Carswell et al. (2004) used osmotic minipumps to deliver either PPT or DPN continuously for 1 week before subjecting mice to 15 min of global ischemia. In this instance, the ERbeta agonist DPN, but not the ERalpha agonist PPT, was found to limit tissue damage.

Taken together the examples outlined above indicate that resveratrol and estrogen confer protection in the same in vivo and in vitro ischemic injury paradigms. Both ERalpha and ERbeta appear to be involved in mediating protection in these contexts. A variety of molecular mechanisms have been proposed to underlie the neuroprotective effects of both resveratrol and estrogen, including actions within mitochondria (Simpkins et al. 2010). Both compounds interact directly with the mitochondrial ATP synthase (Zheng and Ramirez 1999, 2000), stimulate the transcription of mitochondrial respiratory complexes, upregulate the mitochondrial superoxide dismutase enzyme MnSOD, ameliorate mitochondrial reactive oxygen species production and loss of membrane potential while inhibiting the apoptotic pathway (Simpkins et al. 2010).

Parkinson's disease is a neurodegenerative condition characterized by the death of the dopaminergic neurons in the substantia nigra pars compacta, which is involved in coordinating movement. Dietary supplementation with resveratrol confers protection from neurodegeneration in a mouse MPTP model of Parkinson's disease. Pretreatment with resveratrol for 8 days largely prevented the loss of dopamine and tyrosine hydroxylase in striatum, and tyrosine hydroxylase-immunopositive neurons in the substantia nigra following MPTP injection (Blanchet et al. 2008). Similarly, in an alternative model of Parkinson's disease involving direct injection of 6HO-dopamine into the striatum, a 15 day resveratrol pretreatment via dietary supplementation reduced neuronal death and preserved motor functions (Khan et al. 2010). Estrogen is similarly protective against Parkinson's disease. The disorder is more prevalent in men than in women, and women are at greater risk postmenopause as estrogen levels are reduced (reviewed in Bourque et al. 2009). Female mice are also less susceptible than males to MPTP, and methamphetamine (reviewed in Bourque et al. 2009). In both male and ovariectomized female mice, pretreatment with physiological levels of estrogen reduces the extent of injury caused by these neurotoxins. Thus, both estrogen and resveratrol are neuroprotective in similar Parkinson's disease models. Both ERs appear to contribute to Parkinson's disease resistance. Interestingly, ERbeta knockout mice experience degeneration of the substantia nigra in the absence of any exogenous stressor (Wang et al. 2001). In the absence of either ERalpha or ERbeta, estrogen fails to elicit full protection against MPTP injection (Al Sweidi et al. 2012). Interestingly, the ability of resveratrol to upregulate dopamine transporter expression in human dopaminergic neurons is abolished by ER antagonist ICI 182,780, suggesting that some of the relevant effects of resveratrol are exerted via this pathway (Di Liberto et al. 2012).

Thus, there is substantial overlap between the neuroprotective actions of estrogen and resveratrol, while very limited data are available for resveratrol derivatives. In a few instances, the effects of resveratrol have been shown to depend upon an intact ER signalling pathway. However, more directed tests are needed of the hypothesis that the neuroprotection conferred by chronic administration of resveratrol or its derivatives is exerted via ERs. In particular, experiments with resveratrol in ERalpha and ERbeta gene knockout mice will shed light on this hypothesis.

3.4 Mechanisms of Cardioprotective Effects of Resveratrol and Their Relationship to Estrogen Signalling

Dietary resveratrol supplementation can produce a wide variety of beneficial cardiovascular effects in animal models and humans, including protection of heart tissue against ischemic injury. Two weeks of daily intra-gastric resveratrol administration (Dernek et al. 2004) or 1 week of daily resveratrol administration to rats by intraperitoneal injection (Mokni et al. 2007) provides protection against ischemia/reperfusion injury in Langendorff perfused hearts. Recent human trials have also shown improved heart health associated with dietary resveratrol supplementation in post-cardiac infarct patients (Magyar et al. 2012). The beneficial effects of resveratrol on heart function appear to be at least partially attributable to effects on cardiomyocytes per se, as opposed to other cell types. Danz et al. (2009) have shown that cardiomyocytes pre-incubated with 10 μM resveratrol for 72 h are resistant to oxidant-induced injury, perhaps due in part to a substantial upregulation of MnSOD activity that occurs under these conditions. Direct cardioprotective effects of pre-treatment with other red wine polyphenols have not yet been determined.

Similar cardioprotective effects have been associated with estrogens. In many mammalian species, cardiac injury from transient ischemic is less severe in females than males (reviewed in Ostadal et al. 2009), while ovariectomy predisposes female mouse hearts to increased ischemia-reperfusion damage (Nikolic et al. 2007). Effects of ovariectomy in mice can be reversed by 2 weeks of daily estrogen administration, which restores the cardioprotection that is lost in ovariectomized females (Nikolic et al. 2007). ERbeta appears to be involved in estrogen-mediated cardioprotection in mice, since ERbeta$^{-/-}$ mice are more susceptible than wild-type mice to ischemia/reperfusion injury (Gabel et al. 2005; Wang et al. 2008, 2009). Estrogen treatment continues to be cardioprotective in ERalpha$^{-/-}$, but not ERbeta$^{-/-}$, mice (Babiker et al. 2007). Furthermore, DPN administration is as effective as estrogen at protecting the hearts of ovariectomized female mice from ischemia/reperfusion injury (Nikolic et al. 2007).

Dietary supplementation with resveratrol lowers mean arterial blood pressure in humans and animal models of hypertension (see Chap. 2; reviewed in Li et al. 2012).

An important mechanism underlying this effect is the stimulation of endothelial nitric oxide synthase activity and expression, thus enhancing nitric oxide production in the endothelium (see Schini-Kerth et al. 2010 for review). Nitric oxide diffuses out of endothelial cells and into neighboring smooth muscle where it inhibits contraction by a cyclic GMP-mediated pathway. This mechanism of resveratrol's actions in vascular tissue has been investigated using measurement of tension development in isolated aortic rings from a wide variety of species. Grapevine polyphenols elicit relaxation in this model (Fitzpatrick et al. 1993), and the effect is mediated by endothelial cells, since it is absent in vasculature denuded of endothelium. Endothelial nitric oxide synthase activity is affected via two mechanisms: a phosphorylation of serine 1177, and increased protein levels via an induction of transcription. These effects of resveratrol are mediated by ERalpha and ERbeta, and abolished by the ER antagonist ICI 182,780 (Anter et al. 2005; Klinge et al. 2005). Indeed, estrogen exerts virtually identical effects, stimulating endothelial nitric oxide synthase activity via serine 1177 phosphorylation and increased protein synthesis (reviewed in Duckles and Miller 2010). Understanding the role of ER signalling in resveratrol and other wine polyphenols' effects on cardiovascular health is particularly important to understanding the activities of these molecules in both males and females.

3.5 Mechanisms of the Anti-proliferative Effects of Red Wine Polyphenols

A number of reports have linked the growth inhibitory effects of resveratrol to its interactions with ERs (e.g., Bowers et al. 2000; Bhat and Pezzuto 2001; Bhat et al. 2001). Resveratrol's growth inhibitory effects are not restricted to cancer cells, and indeed appear to be rather broad, perhaps dependent upon the relative expression levels of ERalpha and ERbeta, which are generally pro- and anti-proliferative, respectively (Fig. 3.1). Interestingly, the anticancer activities of red wine polyphenols are shared by specific pharmacological agonists of ERbeta. Similar to the anti-proliferative properties of resveratrol, DPN inhibits the growth of the murine colon cancer cell line MC38 in vitro (Motylewska et al. 2009), and of cells in the colon and small intestine of ovariectomized rats in an ERbeta-dependent manner (Schleipen et al. 2011). Loss of ERbeta function is a critical step in the development of prostate cancer (Muthusamy et al. 2011), and breast cancers that express ERbeta generally have a better prognosis than those that do not (Mandusic et al. 2012).

We have shown that resveratrol, pterostilbene, and piceid inhibit proliferation of MRC5 human lung epithelial cells, C2C12 mouse myoblasts, SH-SY5Y human neuroblastoma cells, and primary fibroblast and myoblast cell lines via an ER-dependent mechanism. Further, this effect appears to be mediated by ERbeta and involves mitochondrial redox metabolism as a downstream target (see below for further discussion). Thus, the growth inhibitory effects of resveratrol and related red wine polyphenols may be explained in part by their ability to interact with ERs, in particular ERbeta (Fig. 3.2).

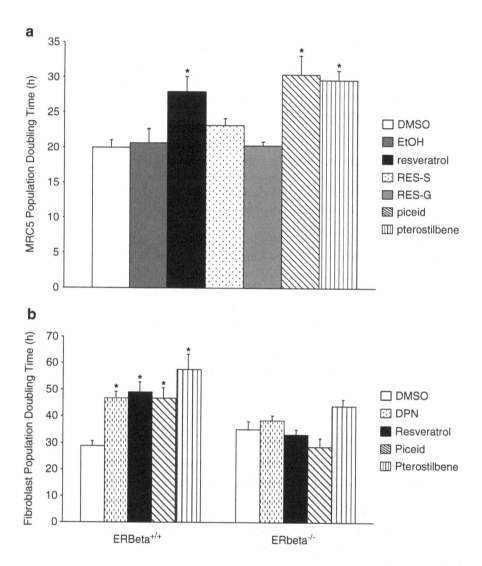

Fig. 3.2 Resveratrol, piceid, and pterostilbene increase population doubling time in an ERbeta-dependent manner. (**a**) Population doubling time in MRC5 fibroblasts treated with DMSO (dimethylsulfoxide; vehicle control), ethanol (EtOH; vehicle control), resveratrol (25 μM), resveratrol-4'-sulfate (RES-S; 50 μM), resveratrol-4'-O-glucuronide (RES-G; 50 μM), piceid (50 μM), or pterostilbene (20 μM) for 48 h ($n = 3$). (**b**) Population doubling time in primary fibroblasts generated from wild-type and ERbeta null mice, treated with DMSO (vehicle control), DPN (ERbeta control; 10 μM), resveratrol (25 μM), piceid (50 μM), or pterostilbene (20 μM) for 48 h ($n = 5$). *Error bars* represent ± SEM, *$p < 0.05$

3.6 Mechanisms Underlying Effects of Grapevine Polyphenols on Mitochondrial Biogenesis and ROS Metabolism

Several of the signalling pathways affected by resveratrol impinge upon mitochondrial functions, including bioenergetics and biogenesis. Lagouge et al. (2006) reported increased mitochondrial abundance, mtDNA copy number and citrate synthase activity in skeletal muscle of high fat diet fed mice given dietary resveratrol supplementation. Similarly, Csiszar et al. (2009) demonstrated increased mitochondrial biogenesis and upregulation of specific mitochondrial proteins in human coronary artery endothelial cells treated with resveratrol. Resveratrol also increases mitochondrial abundance, based on citrate synthase activity in other cell types in vitro, including fibroblasts and myoblasts. Similar results are observed in fibroblasts, myoblasts and prostate cancer cells treated for 24–72 h with pterostilbene or piceid (Fig. 3.3; Robb and Stuart, unpublished data), though no data is yet available for other red wine polyphenols.

Interestingly, virtually identical effects on mitochondrial function have been demonstrated with estrogen (reviewed by Klinge 2008; Chen et al. 2009). For example, estrogen stimulates mitochondrial biogenesis via the transcriptional regulators nuclear respiratory factor-1 and PGC-1alpha. Estrogen administration in vitro upregulates NRF-1 transcription in cerebral blood vessels (Stirone et al. 2005), MCF-7 breast cancer and H1793 lung adenocarcinoma cells (Mattingly et al. 2008). In the latter cells, 4–6d of estrogen treatment also elicited increases in oxygen consumption, and the effect was inhibited by the ERalpha/ERbeta antagonist ICI 182,780. Hsieh et al. (2006) also demonstrated an estrogen-mediated increase in mitochondrial activity in mouse hearts. In vivo, estrogen treatment also increases the expression of multiple mitochondrial genes in brain (reviewed in Brinton 2008), while ovariectomy reduces mitochondrial respiratory chain protein expression and activities in brain tissue (Yao et al. 2010a, b, 2012). Interestingly, estrogen induced mitochondrial biogenesis via PGC-1alpha, and Tfam appears to be required for DPN to confer protection against ischemia/reperfusion injury in rat brain (Hsieh et al. 2006).

Taken together, the results summarized above certainly indicate very similar effects of resveratrol and estrogens on mitochondrial biogenesis. In our experiments, over a 48 h incubation period, estrogen, resveratrol, piceid, pterostilbene, and DPN, but not PPT, all increase apparent mitochondrial abundance based on citrate synthase activity (Fig. 3.4) in several cell lines. Therefore, at least under our experimental conditions, resveratrol and its derivatives appear to stimulate a similar increase in mitochondrial abundance to estrogen.

Resveratrol also targets the mitochondrial antioxidant system in mammalian cells. Resveratrol induces an increase in MnSOD in cardiomyocytes, SK-N-BE neuroblastomas, the HT22 hippocampal neuronal cell line, coronary arterial endothelial cells, PC6.3 pheochromocytoma cells, MRC5 lung fibroblasts, C2C12 myoblasts, SH-SY5Y neuroblastoma, and PC3 prostate cancer cells (Movahed et al. 2012; Albani et al. 2009; Fukui et al. 2010; Ungvari et al. 2009; Kairisalo et al. 2011; Robb and Stuart 2011). In many instances this induction occurs in parallel

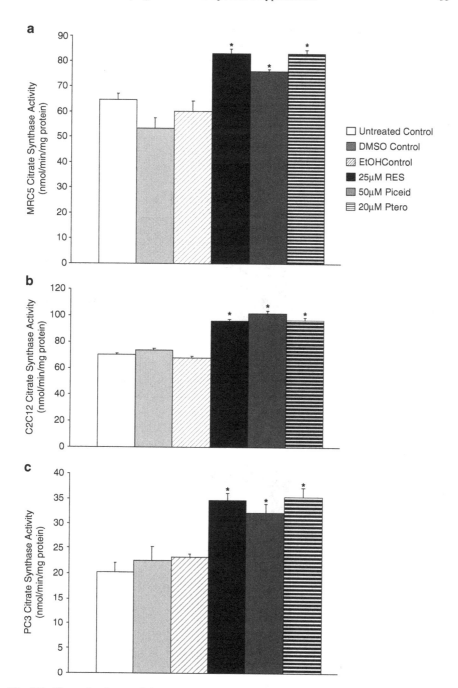

Fig. 3.3 Citrate Synthase activity as a marker of mitochondrial abundance in fibroblasts, C2C12 mouse myoblasts and prostate cancer cells treated with DMSO (vehicle control), ethanol (vehicle control), 20 μM resveratrol, 10 μM pterostilbene, or 50 μM piceid for 48 h. $n = 3$, *error bars* represent ± SEM. *$p < 0.05$

with a slowing or complete inhibition of cell growth, which is interesting given the association between MnSOD activity and mitosis established over the past decade (Sarsour et al. 2012). If the induction of MnSOD elicited by resveratrol is prevented using an siRNA approach, the inhibition of growth is abolished

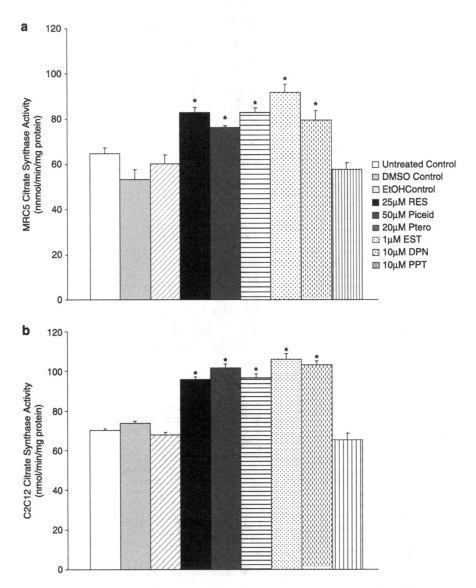

Fig. 3.4 Citrate Synthase activity as a marker of mitochondrial abundance in fibroblasts, myoblasts and prostate cancer cells treated with DMSO (vehicle control), Ethanol (vehicle control), 20 μM resveratrol, 10 μM pterostilbene, 50 μM piceid, 1 μM estradiol, 10 μM DPN, 10 μM PPT for 48 h. $n = 3$, *error bars* represent $\pm$ SEM. *$p < 0.05$

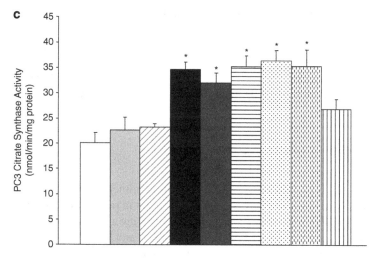

Fig. 3.4 (continued)

(Robb and Stuart 2011). Thus, MnSOD induction appears to play a pivotal role in the ability of resveratrol to regulate mitosis. Using ρ^0 PC3 cells devoid of mitochondrial respiration and therefore producing essentially no matrix superoxide (Hoffmann et al. 2004) we demonstrated the essential role of the superoxide dismutation to hydrogen peroxide in regulating cell growth (Robb and Stuart, unpublished). In ρ^0 cells, although resveratrol treatment elicited a MnSOD induction of similar magnitude to wild type, no effect on growth rate was observed. At this time, the specific redox modification underlying the growth regulation of resveratrol is not known.

MnSOD and mitochondria generally are well-characterized downstream targets of estrogens. In cultured vascular smooth muscle cells estrogen treatment significantly increases MnSOD activity in conjunction with a marked reduction in proliferative cell growth (Sivritas et al. 2011). In rats estrogen treatment increases MnSOD levels in mitochondria isolated from brain tissue (Razmara et al. 2007), and a downregulation of MnSOD is observed in ovariectomized mice (Strehlow et al. 2003). Estrogens also exert rapid and direct effects on MnSOD activity. In mitochondria isolated from breast cancer cells (MCF7) estrogen treatment significantly increases MnSOD activity through an ER-mediated mechanism (Pedram et al. 2006).

We have found that the effects of resveratrol on both MnSOD induction and in turn cell growth are prevented when the ER antagonist ICI 182,780 is included during exposure to resveratrol, indicating that this effect is mediated by estrogen receptors (Robb and Stuart 2011). Since the ERbeta-specific agonist DPN, but not the ERalpha-specific PPT, can phenocopy the growth inhibitory effects of resveratrol, it appears that the MnSOD induction and growth inhibition associated with resveratrol are mediated by ERbeta This hypothesis was confirmed in fibroblasts and myoblasts from ERbeta knockout mice, in which resveratrol fails to induce MnSOD or slow cell growth. Pterostilbene and piceid similarly induce MnSOD expression and slow cell growth (Robb and Stuart, unpublished data). As with resveratrol, when MnSOD induction is prevented using siRNA, ICI 182,780, or ERbeta knockout, neither of these responses occurs. Taken together, these results indicate that the growth inhibitory effects of resveratrol and its derivatives on mammalian cells are due to the induction of MnSOD in an ERbeta dependent manner.

3.7   Anti-obesity Effects

Resveratrol has been termed a "caloric restriction mimetic" based on its ability to elicit some of the same benefits as caloric restriction in rodent models and humans. In an early study of this phenomenon, male mice receiving a high fat diet supplemented with resveratrol showed lower body mass increases and lower percentage body fat than mice receiving just high fat diet, despite having the same overall food intake (Lagouge et al. 2006). These results are concordant with the known effects of estrogens. Estrogen is a major effector of energy homeostasis, body mass, and percentage body fat (reviewed in Faulds et al. 2012). Ovariectomized mice and rats eat more, exercise less and gain more adipose mass than intact controls, effects which can be effectively reversed with regular estrogen treatment (Faulds et al. 2012). Similar effects are typically observed in human females postmenopause (Hirschberg 2012).

Both ERalpha and ERbeta contribute to the effects of estrogen on energy homeostasis: ERalpha-null mice show increases in white adipose tissue mass, and ERbeta-null mice show increases in fatty liver and white adipose tissue mass when fed a high fat diet. ERbeta agonists prevent the increases in body mass, percentage body fat and white adipose tissue mass associated with high fat diet in normal male mice, even though food intake is unaltered (Yepuru et al. 2010). This result is virtually identical to that reported for dietary resveratrol supplementation in mice (Lagouge et al. 2006). Taken together, these results support the hypothesis that resveratrol affects whole body energy homeostasis via ER-mediated signalling.

3.8   Resveratrol as a Sirtuin Activator

Up to this point in our discussion of molecular mechanisms we have focused on emphasizing the many parallels between estrogen signalling and red wine polyphenols. The ability of resveratrol to behave as an ER agonist was first reported in 1997, and for many years this mechanism was hypothesized to account for the pleiotropic effects of this molecule in human health (Gehm et al. 1997; Bowers et al. 2000). Research interest in resveratrol was renewed in 2003, when resveratrol was identified as having antiaging properties in worms and flies (Howitz et al. 2003; Wood et al. 2004). This discovery sparked further investigation into the molecular mechanisms of resveratrol, including the discovery of a controversial interaction with sirtuins that has directed much of the current research into resveratrol's cellular activities.

Sirtuins are a family of highly conserved protein deacetylase enzymes. In yeast, the deacetylase Sir2 was declared a longevity protein following an observation that its overexpression could extend lifespan in *Caenorhabditis elegans* (Tissenbaum and Guarente 2001). Its mammalian orthologue, SIRT1, was thus hypothesized to extend lifespan in mammals. In 2003, Howitz and colleagues identified resveratrol as a potent activator of SIRT1 that increased its reported catalytic activity by approximately 13-fold over control. In yeast, resveratrol similarly increased Sir2 activity, and in agreement with the purported pro-longevity role of Sir2 this was concurrent with a 70 % lifespan extension. Resveratrol was subsequently reported to increase lifespan in *C. elegans, Drosophila sp.* and a short lived vertebrate fish in a supposedly sirtuin-dependent manner (Wood et al. 2004; Valenzano et al. 2006).

In mammals, an activation of SIRT1 activity has been suggested as the molecular mechanism directly responsible for resveratrol's positive effects on metabolism, and cardiovascular and neuronal health. For example, the ability of resveratrol supplementation to increase the health and survival of mice challenged with a high fat diet was postulated to arise from its ability to stimulate sirtuin activity (Baur et al. 2006; Lagouge et al. 2006). The allosteric activation of SIRT1 by resveratrol is thus a prevalent mechanism suggested to account for resveratrol's battery of cellular effects. However, a substantial body of more recent evidence challenges both the hypothesized role for sirtuins in longevity and this putative molecular mechanism of resveratrol.

The evidence to support a direct activation of SIRT1 by resveratrol is at best equivocal, and there are now abundant experimental data inconsistent with this claim. Experiments in which resveratrol treatment increased lifespan in *S. cerevisiae* and *D. melanogaster* in a sirtuin dependent manner similarly could not be replicated, suggesting that the ability of resveratrol to increase lifespan is highly dependent on external factors such as animal husbandry conditions and genetic background (Burnett et al. 2011). In fact, recent studies have failed to corroborate the initial reports of lifespan extension by Sir2 overexpression in *C. elegans* and *D. melanogaster*, concluding that observed effects on lifespan are either smaller than originally reported, strain dependent (Bass et al. 2007) or disappear when the effects of genetic background are adequately controlled for.

Serious questions have also arisen concerning the nature of the in vitro assay used to establish the interaction between resveratrol and SIRT1 (Pacholec et al. 2010; Beher et al. 2009; Borra et al. 2005; Kaeberlein et al. 2005). The initial experiments that identified resveratrol as a direct activator of SIRT1 were based on a deacetylation assay that relied on a fluor-de-lys reporter system. It was later discovered that the fluor-de-lys fluorophore interacts directly with resveratrol resulting in an artificially high signal in the presence of resveratrol (Kaeberlein et al. 2005). Resveratrol does not increase the deacetylase activity of SIRT1 when given its native substrate in the absence of the fluorophore, and thus, data from assays involving the fluor-de-lys fluorophore, detected by both fluorescence and mass spectrometry, are therefore confounded by the artifactually high signal in the presence of resveratrol. A further consideration that challenges the role of a direct interaction between SIRT1 and resveratrol is the discrepancy in plasma and tissue concentrations achievable by dietary dosing regimes and the concentrations shown to increase SIRT1 activity in vitro via allosteric interaction. The concentrations shown to activate SIRT1 by an allosteric interaction in vitro were reported in the micromolar range in a cell free assay using isolated SIRT1 (Howtiz et al. 2003). The conditions of this in vitro experiment do not account for interactions with cell membranes and proteins that substantially affect free concentrations of resveratrol, making the micromolar concentrations required difficult, if not impossible, to achieve in vivo.

There is limited support for a direct interaction with SIRT1 as a mechanistic basis of resveratrol's biological activities. However, SIRT1 has a broad range of molecular targets. For example, acetylation and deacetylation regulate numerous intracellular processes, including various aspects of the cellular stress response (e.g., FOXO transcription factors, heat shock protein factor 1), growth and development (e.g., ERalpha, mammalian target of rapamycin), and cell metabolism (e.g., estrogen-related receptor alpha, PGC1-alpha) (Brunet et al. 2004; Westerheide et al. 2009; Ghosh et al. 2010; Nemoto et al. 2005; Wilson et al. 2010; Yao et al. 2010a, b, c). It is therefore likely that SIRT1 is indirectly involved in resveratrol's cellular effects, since these molecular targets are involved in so many aspects of cell physiology. This diversity of potential targets of SIRT1 may also contribute to the challenge of fully unravelling its role in resveratrol's cellular effects. While experiments in which SIRT1 is deleted have the potential to shed light on the putative SIRT1-resveratrol interaction, deletion of SIRT1 in mammals is extremely harmful, resulting in metabolic deregulation, increased incidence of autoimmune disease, and shortened lifespan (Seifert et al. 2012; Sequeira et al. 2008; Li et al. 2008). Thus, while SIRT1 activity is increased by resveratrol, and deletion of SIRT1 does abolish some of resveratrol's effects on cancer prevention (Boily et al. 2009), this observation is difficult to interpret in light of the extreme phenotype associated with SIRT1 deletion.

In agreement with our argument that many of resveratrol's effects are in fact related to its properties as an ER agonist, the increase in SIRT1 activity reported for resveratrol treatment is phenocopied by estrogen in vivo. In mice, estrogen administration in both old and young ovariectomized mice significantly increased the expression of SIRT1 (Elbaz et al. 2009). Estrogen also stimulates the transcription of SIRT1 in breast cancer cell lines (Elangovan et al. 2011). The ability of resveratrol to activate SIRT1, as has been reported in numerous experimental contexts, may again be reflective of its actions as an ER agonist.

Recently, resveratrol was tested in a conditional SIRT1 knockout that avoided some of the confounding effects of SIRT1 knockout on development. In the absence of SIRT1 many of resveratrol's effects on mitochondrial abundance and respiratory rate in skeletal muscle and adipose tissue were indeed eliminated (Price et al. 2012). However, other resveratrol effects persisted in the absence of SIRT1. Given the importance of sirtuins in biology and the expanding list of proteins whose functions are modified by acetylation/deacetylation, it is likely that signalling pathways regulated by SIRT1 and resveratrol overlap, but the model of a simple, direct activation of SIRT1 activity originally proposed for resveratrol must be abandoned based on evidence accrued to date.

3.9 Resveratrol as an AMP Kinase Activator

Many of resveratrol's purported health benefits relate to its ability to ameliorate the negative effects of high fat, high calorie diets (i.e., reduced insulin sensitivity and hyperlipidemia). On a cellular level an activation of the AMP-activated kinase (AMPK) has been put forward as a mechanism to account for these metabolic effects. AMPK is an important regulator of cellular energy metabolism that is activated by phosphorylation at threonine 172 in response to an increase in the intracellular AMP–ATP ratio, or by the upstream kinases LKB1 and calcium/calmodulin-dependent protein kinase beta. Activation of AMPK occurs in response to a variety of environmental cues including hypoxia, oxidative stress, glucose deprivation and through the actions of hormones relating to energy homeostasis (leptin, adiponectin) (reviewed in Hardie, 2011). Activators of AMPK are used in the treatment of type-2 diabetes, and many of the metabolic effects associated with AMPK activators are phenocopied by resveratrol.

Resveratrol treatment elicits a stimulation of AMPK activity in many cell types both in vitro and in vivo. For example, in cultured human hepatocytes (HepG2 cells) micromolar concentrations of resveratrol elicit a dramatic, and sustained increased in phosphorylated AMPK levels (Zang et al. 2006; Hou et al. 2008; Shin et al. 2009). Resveratrol stimulates AMPK activity in 3T3-L1 adipocytes, an effect that is concurrent with an AMPK dependent inhibition of lipogenesis (Chen et al. 2011). In isolated mouse myotubes (C2C12), a twofold increase in AMPK phosphorylation and elevated glucose uptake are observed with resveratrol treatment (Park et al. 2007). Baur et al. (2006) reported higher levels of phosphorylated AMPK in the liver tissue of male mice supplemented with 400 mg/kg/day resveratrol in a high fat, high calorie diet. Similarly, in an experimental model of obesity, oral administration of 10 mg/kg/day resveratrol ameliorated dyslipidemia and insulin resistance in Zuker rats. These positive changes in metabolic health were associated with an increase in AMPK phosphorylation in liver tissue (Rivera et al. 2009). In skeletal muscle, an important site of glucose metabolism, dietary resveratrol increases the phosphorylation of AMPK in a SIRT1 dependent manner (Price et al. 2012).

Evidence for AMPK activation in humans given a supplemental dose of resveratrol is limited due to the low number of clinical trials that have been completed to date. In healthy, but obese men, 30 days of a 150 mg/day resveratrol supplement resulted in elevated levels of phosphorylated AMPK in skeletal muscle tissue (Timmers et al. 2011). However, in a 12-week trial of postmenopausal women a 75 mg/day dose of resveratrol failed to evoke a significant change in phosphorylated AMPK levels in skeletal muscle or adipose tissue (Yoshino et al. 2012). More data is required to fully appreciate the relationship between AMPK and resveratrol in humans.

AMPK activation has also been implicated in the neuroprotective effects of resveratrol. An activation of AMPK in primary neurons was first reported by Dasgupta and Milbrandt in 2007, who demonstrated that resveratrol treatment stimulated AMPK phosphorylation and activation of its downstream targets in Neuro2a neuroblastoma cells. Resveratrol stimulates differentiation of the Neuro2a cells, and using a dominant negative of AMPK it was determined that a functional AMPK is required for this biological activity. In 2 month old male mice intraperitoneal injection of 20 mg/kg resveratrol stimulates an increase in phosphorylated AMPK in brain tissue (Dasgupta and Milbrandt 2007). Similarly, oral administration of resveratrol for 15 weeks significantly increases levels of phosphorylated AMPK in the brain tissue of mice, an effect that is associated with reduced levels of amyloid beta peptide (Vingtdeux et al. 2010).

The activation of AMPK has also been cited as the mechanism responsible for resveratrol`s inhibition of cell growth. In neural progenitor cells isolated from mice RES treatment significantly reduces proliferation in an AMPK dependent manner (Park et al. 2012). Fourteen days of resveratrol treatment in mice significantly activates AMPK, and reduces the proliferation and survival of neural progenitor cells in the dentate gyrus of the hippocampus, which manifests at the organism level as a reduction in spatial learning and memory capacity (Park et al. 2012). Resveratrol activation of AMPK is also associated with the inhibition of proliferative cell growth in breast cancer cell lines (MDA-MB-231 and MCF7) (Lin et al. 2010), and in cardiomyocytes which is hypothesized to confer protection against cardiac hypertrophy (Chan et al. 2008; Hwang et al. 2008).

There is relatively little data regarding the effect of other red wine polyphenols on AMPK activation. However, similar to resveratrol pterostilbene stimulates AMPK phosphorylation in prostate cancer cell lines (PC3, LNCaP) and suppresses proliferation (Lin et al. 2010, 2012). The inhibition of proliferation arising from pterostilbene treatment can be prevented with a pharmacological inhibitor of AMPK (Lin et al. 2012), which demonstrates that as is the case with resveratrol, this effect requires AMPK activity. A 5 μM mixture of resveratrol, quercetin, and catechin, increased AMPK activity in MD-MBA-231 breast cancer cell lines, and in vivo a 5 mg/kg/day dose of these compounds for 1 week effectively reduces the growth mammary fat pad tumors in an AMPK-dependent manner (Castillo-Pichardo and Dharmawardhane 2012).

AMPK activation is observed in, and required for many of the potentially positive effects of resveratrol and red wine polyphenols on human health (metabolic effects, neuroprotection, and anticancer properties). However, the mechanism responsible for the AMPK activation is contentious. Activation of AMPK by resveratrol is not due to a direct interaction, as no direct stimulation of AMPK activity is observed in a cell free, in vitro assay (Baur et al. 2006). It also appears that the activation of AMPK by resveratrol is not a response to a change in intracellular AMP concentrations. In cultured neurons no significant change in the ratio of AMP–ATP concentrations was detected in response to resveratrol treatment (Dasgupta and Milbrandt, 2007), suggesting that AMPK phosphorylation is downstream of another resveratrol stimulated pathway. Activation of AMPK through SIRT1 dependent and independent mechanisms have been reported (e.g., Price et al.

2012; Dasgupta and Milbrandt 2007), but given the importance of both SIRT1 and AMPK to metabolic health experiments involving the deletion or knockdown of these proteins are complicated by the overall severity of the resulting phenotype.

An often overlooked but important connection is the relationship between estrogen and AMPK activity. Estrogen is a key regulator of energy homeostasis (reviewed in Faulds et al. 2012) and, similar to resveratrol, stimulates the activity of AMPK in cultured cells in vitro and in tissues in vivo. In mouse myoblasts (C2C12 cells) estradiol elicits a significant and rapid increase in AMPK phosphorylation (D'Eon et al. 2008). Estradiol also increases AMPK phosphorylation in 3T3-L1 adipocytes, an effect that is inhibited by the ER antagonist ICI182,780 (Kim et al. 2012). In vivo, a fivefold induction of phosphorylated AMPK in muscle tissue is observed in response to estradiol treatment of ovariectomized mice. The rapid activation of AMPK by estradiol is abolished by an ER antagonist in cultured myoblasts (D'Eon et al. 2005). An activation of AMPK is observed in skeletal muscle of ovariectomized rats given PPT, but not DPN or oestradiol benzoate for 3 days (Gorres et al. 2011), which is suggestive of a critical role for ERalpha, but not ERbeta in this response. In freshly isolated rat soleus muscle 10 nM estradiol rapidly activates AMPK in an ER-dependent manner (Rogers et al. 2009). Thus, estrogen is also a potent activator of AMPK both in vitro and in vivo. Indeed, many of the reported effects of red wine polyphenols on AMPK activity are very similar to those that have been associated with estrogen. Despite this, the contribution of ERs to the activation of AMPK by resveratrol and its derivatives remain surprisingly unexplored. In our opinion, it seems highly plausible that resveratrol's effects on AMPK are at least partially mediated by ER signalling pathways.

3.10   Effects of Resveratrol on Phosphodiesterase and AMP

In a recent publication, Park et al. (2012) reported another putative molecular mechanism associated with resveratrol as an inhibitor of phosphodiesterase activity. These authors reported a dramatic increase in cAMP levels in C2C12 myotubes treated with 10 µM resveratrol. cAMP levels may be increased by either a stimulation of adenylate cyclase activity (the enzyme responsible for cAMP synthesis), or an inhibition of phosphodiesterase activity (which catalyzes cAMP degradation). Park et al. (2012) demonstrated that the metabolic effects of resveratrol in cultured myotubes were independent of adenylate cyclase activity, but could be replicated by rolipram, an inhibitor of phosphodiesterase activity. Using an in vitro assay of phosphodiesterase activity they reported that resveratrol acted as a competitive inhibitor of phosphodiesterase isoforms 1, 3, and 4. In vivo, the metabolic effects of resveratrol supplementation could be phenocopied with rolipram (Park et al. 2012).

This study was not the first to describe a role for cAMP in resveratrol's molecular mechanism. Resveratrol has been shown to increase cAMP levels in MCF7 breast cancer cells, and rolipram can reproduce resveratrol's anti-proliferative activities in these cells (El-Mowafy and Alkhalaf 2003). However, in this study the authors present evidence that resveratrol is activating adenylate cyclase, and not inhibiting phosphodiesterase activity (El-Mowafy and Alkhalaf 2003). In another study, low micromolar concentrations of resveratrol and piceid had negligible effects on phosphodiesterase (isoform 5) activity (Dell'Agli et al. 2005).

Similar to resveratrol, estrogen also appears to alter intracellular cyclic AMP levels, but this may be via a different mechanism than that observed for resveratrol above. In cultured differentiated monocytic leukemia cells (THP-1) estrogen treatment increases cAMP concentrations via a stimulation of adenylate cylcase activity, but has no effect on phosphodiesterase activity (Kanda and Watanabe 2002). An increase in cAMP concentrations is also observed in rabbit proximal tubule cells treated with estrogen (Han et al. 2000). However, the available evidence suggests the effects of estrogen on cAMP levels are via adenylate cyclase activation rather than phosphodiesterase inhibition (e.g., Filardo et al. 2002). More research is needed to understand the potential contribution of phosphodiesterase inhibition to the actions of resveratrol and its derivatives in mammalian cells.

3.11   Conclusions

The molecular mechanisms responsible for the effects of resveratrol and its derivatives on human health continue to be the subject of much debate within the scientific community. Although a wide range of specific inter-molecular interactions have been proposed for resveratrol, it is interesting to note the striking parallels that exist between the effects of resveratrol and those of estrogens. Many of the signalling pathways hypothesized to account for resveratrol's actions converge with estrogen signalling pathways. The possibility that resveratrol's estrogenic properties underlie these purported mechanisms (as outlined in Fig. 3.5) has not been fully investigated,

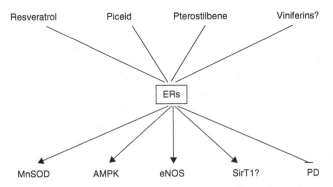

Fig. 3.5 In mammals, resveratrol and its derivatives in grapevine may exert similar effects on a variety of downstream cellular targets via estrogen receptor-mediated signalling. *ERs* estrogen receptors, *MnSOD* Mn superoxide dismutase, *AMPK* AMP kinase, *eNOS* endothelial nitric oxide synthase, *SIRT1* sirtuin isoform 1, *PD* phosphodiesterase

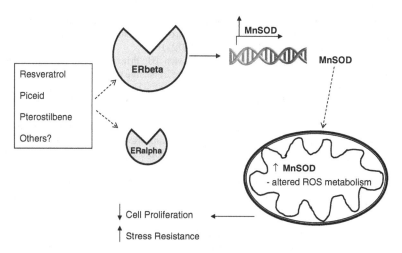

Fig. 3.6 Resveratrol and derivatives affect cellular stress resistance and proliferative growth via primarily ERbeta-mediated modulation of mitochondrial ROS metabolism

and more resveratrol experiments should be undertaken in ER-null (particularly ERbeta-null) mice.

In the expanse of data collected to date describing resveratrol and other red wine polyphenol's cellular activities, mitochondria have emerged as a clear proximal target of the actions of these molecules (see, e.g., Fig. 3.6). Indeed, mitochondrial biogenesis and altered ROS metabolism are common observations made with both resveratrol and estrogen treatment in vitro and in vivo. Understanding how these changes in mitochondrial function manifest on an organismal level to yield the positive effects observed with these molecules will provide significant insight into human health research.

References

Agrawal M, Kumar V, Kashyap MP, Khanna VK, Randhawa GS, Pant AB (2011) Ischemic insult induced apoptotic changes in PC12 cells: protection by trans resveratrol. Eur J Pharmacol 666:5–11

Aguirre CC, Baudry M (2009) Progesterone reverses 17β-estradiol-mediated neuroprotection and BDNF induction in cultured hippocampal slices. Eur J Neurosci 29:447–454

Aguirre C, Jayaraman A, Pike C, Baudry M (2010) Progesterone inhibits estrogen-mediated neuroprotection against excitotoxicity by down-regulating estrogen receptor-β. J Neurochem 115:1277–1287

Albani D, Polito L, Batelli S, De Mauro S, Fracasso C, Martelli G, Colombo L, Manzoni C, Salmona M, Caccia S, Negro A, Forloni G (2009) The SIRT1 activator resveratrol protects SK-N-BE cells from oxidative stress and against toxicity caused by alpha-synuclein or amyloid-beta (1-42) peptide. J Neurochem 2110(5):1445–1456

Anter E, Chen K, Shapira OM, Karas RH, Keaney JF Jr (2005) p38 mitogen-activated protein kinase activates eNOS in endothelial cells by an estrogen receptor alpha-dependent pathway in response to black tea polyphenols. Circ Res 96:1072–1078

Babiker FA, Lips DJ, Delvaux E, Zandberg P, Janssen BJ, Prinzen F, van Eys G, Grohé C, Doevendans PA (2007) Oestrogen modulates cardiac ischaemic remodelling through oestrogen receptor-specific mechanisms. Acta Physiol (Oxf) 189:23–31

Barros RP, Gustafsson JÅ (2011) Estrogen receptors and the metabolic network. Cell Metab 14:289–299

Bass TM, Weinkove D, Houthoofd K, Gems D, Partridge L (2007) Effects of resveratrol on lifespan in *Drospholia melanogaster* and *Caenorhabditis elegans*. Mech Ageing Dev 128:546–552

Baur JA, Pearson KJ, Price NL, Jamieson HA, Lerin C, Kalra A, Prabhu VV, Allard JS, Lopez-Lluch G, Lewis K, Pistell PJ, Poosala S, Becker KG, Boss O, Gwinn D, Wang M, Ramaswamy S, Fishbein KW, Spencer RG, Lakatta EG, Le Couteur D, Shaw RJ, Navas P, Puigserver P, Ingram DK, de Cabo R, Sinclair DA (2006) Resveratrol improves health and survival of mice on a high-calorie diet. Nature 444:337–342

Beher D, Wu J, Cumine S, Kim KW, Lu SC, Atangan L, Wang M (2009) Resveratrol is not a direct activator of SIRT1 enzyme activity. Chem Biol Drug Des 74:619–624

Bhat KP, Pezzuto JM (2001) Resveratrol exhibits cytostatic and antiestrogenic properties with human endometrial adenocarcinoma (Ishikawa) cells. Cancer Res 61:6137–6144

Bhat KP, Lantvit D, Christov K, Mehta RG, Moon RC, Pezzuto JM (2001) Estrogenic and antiestrogenic properties of resveratrol in mammary tumor models. Cancer Res 61:7456–7463

Blanchet J, Longpré F, Bureau G, Morissette M, DiPaolo T, Bronchti G, Martinoli MG (2008) Resveratrol, a red wine polyphenol, protects dopaminergic neurons in MPTP-treated mice. Prog Neuropsychopharmacol Biol Psychiatry 32:1243–1250

Boily G, He XH, Pearce B, Jardine K, McBurney MW (2009) SirT1-null mice develop tumors at normal rates but are poorly protected by resveratrol. Oncogene 28:2882–2893

Borra MT, Smith BC, Denu JM (2005) Mechanism of human SIRT1 activation by resveratrol. J Biol Chem 280:17187–17195

Bourque M, Dluzen DE, Di Paolo T (2009) Neuroprotective actions of sex steroids in Parkinson's disease. Front Neuroendocrinol 30:142–157

Bowers JL, Tyulmenkov VV, Jernigan SC, Klinge CM (2000) Resveratrol acts as a mixed agonist/antagonist for estrogen receptors alpha and beta. Endocrinology 141:3657–3667

Brandenberger AW, Tee MK, Lee JY, Chao V, Jaffe RB (1997) Tissue distribution of estrogen receptors alpha (ER-alpha) and beta (ER-beta) mRNA in the midgestational human fetus. J Clin Endocrinol Metab 82(10):3509–3512

Brinton RD (2008) Estrogen regulation of glucose metabolism and mitochondrial function: therapeutic implications for prevention of Alzheimer's disease. Adv Drug Deliv Rev 60:1504–1511

Brunet A, Sweeney LB, Sturgill JF, Chua KF, Greer PL, Lin Y, Tran H, Ross SE, Mostoslavsky R, Cohen HY, Hu LS, Cheng HL, Jedrychowski MP, Gygi SP, Sinclair DA, Alt FW, Greenberg ME (2004) Stress-dependent regulation of FOXO transcription factors by the SIRT1 deacetylase. Science 303:2011–2015

Burnett C, Valentini S, Cabreiro F, Goss M, Somogyvári M, Piper MD, Hoddinott M, Sutphin GL, Leko V, McElwee JJ, Vazquez-Manrique RP, Orfila AM, Ackerman D, Au C, Vinti G, Riesen M, Howard K, Neri C, Bedalov A, Kaeberlein M, Soti C, Partridge L, Gems D (2011) Absence of effects of Sir2 overexpression on lifespan in C. elegans and Drosophila. Nature 477:482–485

Carswell HV, Macrae IM, Gallagher L, Harrop E, Horsburgh KJ (2004) Neuroprotection by a selective estrogen receptor beta agonist in a mouse model of global ischemia. Am J Physiol Heart Circ Physiol 287:H1501–H1504

Castillo-Pichardo L, Dharmawardhane SF (2012) Grape polyphenols inhibit akt/mammalian target of rapamycin signaling and potentiate the effects of gefitinib in breast cancer. Nutr Cancer 64:1058–1069

Chan AY, Dolinsky VW, Soltys CL, Viollet B, Baksh S, Light PE, Dyck JR (2008) Resveratrol inhibits cardiac hypertrophy via AMP-activated protein kinase and Akt. J Biol Chem 283:24194–24201

Chen JQ, Cammarata PR, Baines CP, Yager JD (2009) Regulation of mitochondrial respiratory chain biogenesis by estrogens/estrogen receptors and physiological, pathological and pharmacological implications. Biochim Biophys Acta 1793:1540–1570

Chen S, Li Z, Li W, Shan Z, Zhu W (2011) Resveratrol inhibits cell differentiation in 3T3-L1 adipocytes via activation of AMPK. Can J Physiol Pharmacol 89:793–799

Cheng Y, Zhang HT, Sun L, Guo S, Ouyang S, Zhang Y, Xu J (2006) Involvement of cell adhesion molecules in polydatin protection of brain tissues from ischemia-reperfusion injury. Brain Res 1110:193–200

Csiszar A, Labinskyy N, Pinto JT, Ballabh P, Zhang H, Losonczy G, Pearson K, de Cabo R, Pacher P, Zhang C, Ungvari Z (2009) Resveratrol induces mitochondrial biogenesis in endothelial cells. Am J Physiol Heart Circ Physiol 297:H13–H20

Danz ED, Skramsted J, Henry N, Bennett JA, Keller RS (2009) Resveratrol prevents doxorubicin cardiotoxicity through mitochondrial stabilization and the Sirt1 pathway. Free Radic Biol Med 46:1589–1597

Dasgupta B, Milbrandt J (2007) Resveratrol stimulates AMP kinase activity in neurons. Proc Natl Acad Sci USA 104:7217–7222

Della-Morte D, Dave KR, DeFazio RA, Bao YC, Raval AP, Perez-Pinzon MA (2009) Resveratrol pretreatment protects rat brain from cerebral ischemic damage via a sirtuin 1-uncoupling protein 2 pathway. Neuroscience 159:993–1002

D'Eon TM, Souza SC, Aronovitz M, Obin MS, Fried SK, Greenberg AS (2005) Estrogen regulation of adiposity and fuel partitioning. Evidence of genomic and non-genomic regulation of lipogenic and oxidative pathways. J Biol Chem 280:35983–35991

D'Eon TM, Rogers NH, Stancheva ZS, Greenberg AS (2008) Estradiol and the estradiol metabolite, 2-hydroxyestradiol, activate AMP-activated protein kinase in C2C12 myotubes. Obesity (Silver Spring) 16:1284–1288

Dell'Agli M, Galli GV, Vrhovsek U, Mattivi F, Bosisio E (2005) In vitro inhibition of human cGMP-specific phosphodiesterase-5 by polyphenols from red grapes. J Agric Food Chem 53(6):1960–1965

Dernek S, Ikizler M, Erkasap N, Ergun B, Koken T, Yilmaz K, Sevin B, Kaygisiz Z, Kural T (2004) Cardioprotection with resveratrol pretreatment: improved beneficial effects over standard treatment in rat hearts after global ischemia. Scand Cardiovasc J 38:245–254

Di Liberto V, Mäkelä J, Korhonen L, Olivieri M, Tselykh T, Mälkiä A, Do Thi H, Belluardo N, Lindholm D, Mudò G (2012) Involvement of estrogen receptors in the resveratrol-mediated increase in dopamine transporter in human dopaminergic neurons and in striatum of female mice. Neuropharmacology 62:1011–1018

Duckles SP, Miller VM (2010) Hormonal modulation of endothelial NO production. Pflugers Arch – EurJ Physiol 459:841–851

El-Mowafy AM, Alkhalaf M (2003) Resveratrol activates adenylyl-cyclase in human breast cancer cells: a novel, estrogen receptor-independent cytostatic mechanism. Carcinogenesis 24(5):869–873

Elangovan S, Ramachandran S, Venkatesan N, Ananth S, Gnana-Prakasam JP, Martin PM, Browning DD, Schoenlein PV, Prasad PD, Ganapathy V, Thangaraju M (2011) SIRT1 is essential for oncogenic signaling by estrogen/estrogen receptor α in breast cancer. Cancer Res 71(21):6654–6664

Elbaz A, Rivas D, Duque G (2009) Effect of estrogens on bone marrow adipogenesis and Sirt1 in aging C57BL/6J mice. Biogerontology 10(6):747–755

Faulds MH, Zhao C, Dahlman-Wright K, Gustafsson JÅ (2012) The diversity of sex steroid action: regulation of metabolism by estrogen signaling. J Endocrinol 212:3–12

Filardo EJ, Quinn JA, Frackelton AR Jr, Bland KI (2002) Estrogen action via the G protein-coupled receptor, GPR30: stimulation of adenylyl cyclase and cAMP-mediated attenuation of the epidermal growth factor receptor-to-MAPK signaling axis. Mol Endocrinol 16(1):70–84

Fitzpatrick DF, Hirschfield SL, Coffey RG (1993) Effects of red and white wine on endothelium-dependent vasorelaxation of rat aort and human coronary arteries. Am J Physiol 275:H1183–H1190

Fukui M, Choi HJ, Zhu BT (2010) Mechanism for the protective effect of resveratrol against oxidative stress-induced neuronal death. Free Radic Biol Med 49(5):800–813

Gabel SA, Walker VR, London RE, Steenbergen C, Korach KS, Murphy E (2005) Estrogen receptor beta mediates gender differences in ischemia/reperfusion injury. J Mol Cell Cardiol 38:289–297

Gehm BD, McAndrews JM, Chien PY, Jameson JL (1997) Resveratrol, a polyphenolic compound found in grapes and wine, is an agonist for the estrogen receptor. Proc Natl Acad Sci USA 94:14138–14143

Ghosh HS, McBurney M, Robbins PD (2010) SIRT1 negatively regulates the mammalian target of rapamycin. PLoS One 5(2):e9199

Gorres BK, Bomhoff GL, Morris JK, Geiger PC (2011) In vivo stimulation of oestrogen receptor α increases insulin-stimulated skeletal muscle glucose uptake. J Physiol 589:2041–2054

Guo J, Krause DN, Horne J, Weiss JH, Li X, Duckles SP (2010) Estrogen-receptor-mediated protection of cerebral endothelial cell viability and mitochondrial function after ischemic insult in vitro. J Cereb Blood Flow Metab 30:545–554

Guo J, Duckles SP, Weiss JH, Li X, Krause DN (2012) 17β-Estradiol prevents cell death and mitochondrial dysfunction by an estrogen receptor-dependent mechanism in astrocytes after oxygen-glucose deprivation/reperfusion. Free Radic Biol Med 52:2151–2160

Han HJ, Lee YH, Park SH (2000) Estradiol-17beta-BSA stimulates Ca(2+) uptake through non-genomic pathways in primary rabbit kidney proximal tubule cells: involvement of cAMP and PKC. J Cell Physiol 183(1):37–44

Han S, Choi JR, Soon Shin K, Kang SJ (2012) Resveratrol upregulated heat shock proteins and extended the survival of G93A-SOD1 mice. Brain Res 5(1483):112–117

Hardie DG (2011) AMP-activated protein kinase: an energy sensor that regulates all aspects of cell function. Genes Dev 25:1895–1908

Hirschberg AL (2012) Sex hormones, appetite and eating behaviour in women. Maturitas 71:248–256

Hoffmann S, Spitkovsky D, Radicella JP, Epe B, Wiesner RJ (2004) Reactive oxygen species derived from the mitochondrial respiratory chain are not responsible for the basal levels of oxidative base modifications observed in nuclear DNA of Mammalian cells. Free Radic Biol Med 36(6):765–773

Horsburgh K, Macrae IM, Carswell H (2002) Estrogen is neuroprotective via an apolipoprotein E-dependent mechanism in a mouse model of global ischemia. J Cereb Blood Flow Metab 22:1189–1195

Hou X, Xu S, Maitland-Toolan KA, Sato K, Jiang B, Ido Y, Lan F, Walsh K, Wierzbicki M, Verbeuren TJ, Cohen RA, Zang M (2008) SIRT1 regulates hepatocyte lipid metabolism through activating AMP-activated protein kinase. J Biol Chem 283:20015–20026

Howitz KT, Bitterman KJ, Cohen HY, Lamming DW, Lavu S, Wood JG, Zipkin RE, Chung P, Kisielewski A, Zhang LL, Scherer B, Sinclair DA (2003) Small molecule activators of sirtuins extend Saccharomyces cerevisiae lifespan. Nature 425(6954):191–196

Hsieh YC, Choudhry MA, Yu HP, Shimizu T, Yang S, Suzuki T, Chen J, Bland KI, Chaudry IH (2006) Inhibition of cardiac PGC-1alpha expression abolishes ERbeta agonist-mediated cardioprotection following trauma-hemorrhage. FASEB J 20:1109–1117

Hu Y, Liu J, Wang J, Liu Q (2011) The controversial links among calorie restriction, SIRT1, and resveratrol. Free Radic Biol Med 2011(51):250–256

Hwang JT, Kwon DY, Park OJ, Kim MS (2008) Resveratrol protects ROS-induced cell death by activating AMPK in H9c2 cardiac muscle cells. Genes Nutr 2(4):323–326

Ji H, Zhang X, Du Y, Liu H, Li S, Li L (2012) Polydatin modulates inflammation by decreasing NF-κB activation and oxidative stress by increasing Gli1, Ptch1, SOD1 expression and ameliorates blood-brain barrier permeability for its neuroprotective effect in pMCAO rat brain. Brain Res Bull 87:50–59

Kaeberlein M, McDonagh T, Heltweg B, Hixon J, Westman EA, Caldwell SD, Napper A, Curtis R, DiStefano PS, Fields S, Bedalov A, Kennedy BK (2005) Substrate-specific activation of sirtuins by resveratrol. J Biol Chem 280:17038–17045

Kairisalo M, Bonomo A, Hyrskyluoto A, Mudò G, Belluardo N, Korhonen L, Lindholm D (2011) Resveratrol reduces oxidative stress and cell death and increases mitochondrial antioxidants and XIAP in PC6.3-cells. Neurosci Lett 488(3):263–266

Kanda N, Watanabe S (2002) 17beta-estradiol enhances vascular endothelial growth factor production and dihydrotestosterone antagonizes the enhancement via the regulation of adenylate cyclase in differentiated THP-1 cells. J Invest Dermatol 118(3):519–529

Khan MM, Ahmad A, Ishrat T, Khan MB, Hoda MN, Khuwaja G, Raza SS, Khan A, Javed H, Vaibhav K, Islam F (2010) Resveratrol attenuates 6-hydroxydopamine-induced oxidative damage and dopamine depletion in rat model of Parkinson's disease. Brain Res 1328:139–151

Kim JY, Jo KJ, Kim BJ, Baik HW, Lee SK (2012) 17β-estradiol induces an interaction between adenosine monophosphate-activated protein kinase and the insulin signaling pathway in 3T3-L1 adipocytes. Int J Mol Med, In Press

Klinge CM, Blankenship KA, Risinger KE, Bhatnagar S, Noisin EL, Sumanasekera WK, Zhao L, Brey DM, Keynton RS (2005) Resveratrol and estradiol rapidly activate MAPK signaling through estrogen receptors alpha and beta in endothelial cells. J Biol Chem 280:7460–7468

Klinge CM (2008) Estrogenic control of mitochondrial function and biogenesis. J Cell Biochem 105(6):1342–1351

Lagouge M, Argmann C, Gerhart-Hines Z, Meziane H, Lerin C, Daussin F, Messadeq N, Milne J, Lambert P, Elliott P, Geny B, Laakso M, Puigserver P, Auwerx J (2006) Resveratrol improves mitochondrial function and protects against metabolic disease by activating SIRT1 and PGC-1alpha. Cell 127(6):1109–1122

Leitman DC, Paruthiyil S, Vivar OI, Saunier EF, Herber CB, Cohen I, Tagliaferri M, Speed TP (2010) Regulation of specific target genes and biological responses by estrogen receptor subtype agonists. Curr Opin Pharmacol 10(6):629–636

Li H, Xia N, Förstermann U. (2012) Cardiovascular effects and molecular targets of resveratrol. Nitric Oxide 26:102–110

Li Y, Xu W, McBurney MW, Longo VD (2008) SirT1 inhibition reduces IGF-I/IRS-2/Ras/ERK1/2 signaling and protects neurons. Cell Metab 8:38–48

Lin JN, Lin VC, Rau KM, Shieh PC, Kuo DH, Shieh JC, Chen WJ, Tsai SC, Way TD (2010) Resveratrol modulates tumor cell proliferation and protein translation via SIRT1-dependent AMPK activation. J Agric Food Chem 58:1584–1592

Lin VC, Tsai YC, Lin JN, Fan LL, Pan MH, Ho CT, Wu JY, Way TD (2012) Activation of AMPK by pterostilbene suppresses lipogenesis and cell-cycle progression in p53 positive and negative human prostate cancer cells. J Agric Food Chem 60:6399–6407

Magyar K, Halmosi R, Palfi A, Feher G, Czopf L, Fulop A, Battyany I, Sumegi B, Toth K, Szabados E (2012) Cardioprotection by resveratrol: a human clinical trial in patients with stable coronary artery disease. Clin Hemorheol Microcirc 50:179–187

Mandusic V, Dimitrijevic B, Nikolic-Vukosavljevic D, Neskovic-Konstantinovic Z, Kanjer K, Hamann U (2012) Different associations of estrogen receptor β isoforms, ERβ1 and ERβ2, expression levels with tumor size and survival in early- and late-onset breast cancer. Cancer Lett 321(1):73–79

Mattingly KA, Ivanova MM, Riggs KA, Wickramasinghe NS, Barch MJ, Klinge CM (2008) Estradiol stimulates transcription of nuclear respiratory factor-1 and increases mitochondrial biogenesis. Mol Endocrinol 22:609–622

Movahed A, Yu L, Thandapilly SJ, Louis XL, Netticadan T (2012) Resveratrol protects adult cardiomyocytes against oxidative stress mediated cell injury. Arch Biochem Biophys 527(2):74–80

Mokni M, Limam F, Elkahoui S, Amri M, Aouani E (2007) Strong cardioprotective effect of resveratrol, a red wine polyphenol, on isolated rat hearts after ischemia/reperfusion injury. Arch Biochem Biophys 457:1–6

Motylewska E, Stasikowska O, Mełeń-Mucha G (2009) The inhibitory effect of diarylpropionitrile, a selective agonist of estrogen receptor beta, on the growth of MC38 colon cancer line. Cancer Lett 276(1):68–73

Muthusamy S, Andersson S, Kim HJ, Butler R, Waage L, Bergerheim U, Gustafsson JÅ (2011) Estrogen receptor β and 17β-hydroxysteroid dehydrogenase type 6, a growth regulatory pathway that is lost in prostate cancer. Proc Natl Acad Sci USA 108(50):20090–20094

Nemoto S, Fergusson MM, Finkel T (2005) SIRT1 functionally interacts with the metabolic regulator and transcriptional coactivator PGC-1{alpha}. J Biol Chem 280:16456–16460

Nikolic I, Liu D, Bell JA, Collins J, Steenbergen C, Murphy E (2007) Treatment with an estrogen receptor-beta-selective agonist is cardioprotective. J Mol Cell Cardiol 42:769–780

Nilsson S, Koehler KF, Gustafsson JÅ (2011) Development of subtype-selective oestrogen receptor-based therapeutics. Nat Rev Drug Discov 10:778–792

Ostadal B, Netuka I, Maly J, Besik J, Ostadalova I (2009) Gender differences in cardiac ischemic injury and protection-experimental aspects. Exp Biol Med 234:1011–1019

Pacholec M, Bleasdale JE, Chrunyk B, Cunningham D, Flynn D, Garofalo RS, Griffith D, Griffor M, Loulakis P, Pabst B, Qiu X, Stockman B, Thanabal V, Varghese A, Ward J, Withka J, Ahn K (2010) SRT1720, SRT2183, SRT1460, and resveratrol are not direct activators of SIRT1. J Biol Chem 285(11):8340–8351

Park CE, Kim MJ, Lee JH, Min BI, Bae H, Choe W, Kim SS, Ha J (2007) Resveratrol stimulates glucose transport in C2C12 myotubes by activating AMP-activated protein kinase. Exp Mol Med 39:222–229

Park HR, Kong KH, Yu BP, Mattson MP, Lee JJ (2012) Resveratrol inhibits the proliferation of neural progenitor cells and hippocampal neurogenesis. Biol Chem 287:42588–42600

Pedram A, Razandi M, Wallace DC, Levin ER (2006) Functional estrogen receptors in the mitochondria of breast cancer cells. Mol Biol Cell 17(5):2125–2137

Price NL, Gomes AP, Ling AJY, Duarte FV, Martin-Montalvo A, North BJ, Agarwal B, Ye L, Ramadori G, Teodoro JS, Hubbard BP, Varela AT, Davis JG, Varamini B, Hafner A, Moaddel R, Rolo AP, Coppari R, Palmeira CM, de Cabo R, Baur JA, Sinclair DA (2012) SIRT1 is required for AMPK activation and the beneficial effects of resveratrol on mitochondrial function. Cell Metab 2012(15):675–690

Razmara A, Duckles SP, Krause DN, Procaccio V (2007) Estrogen suppresses brain mitochondrial oxidative stress in female and male rats. Brain Res 1176:71–81

Raval AP, Dave KR, Pérez-Pinzón MA (2006) Resveratrol mimics ischemic preconditioning in the brain. J Cereb Blood Flow Metab 26:1141–1147

Ren J, Fan C, Chen N, Huang J, Yang Q (2011) Resveratrol pretreatment attenuates cerebral ischemic injury by upregulating expression of transcription factor Nrf2 and HO-1 in rats. Neurochem Res 36:2352–2362

Rivera L, Moró R, Zarzuelo A, Galisteo M (2009) Long-term resveratrol administration reduces metabolic disturbances and lowers blood pressure in obese Zucker rats. Biochem Pharmacol 77:1053–1063

Robb EL, Page MM, Wiens BE, Stuart JA (2008a) Molecular mechanisms of oxidative stress resistance induced by resveratrol: specific and progressive induction of MnSOD. Biochem Biophys Res Commun 367:406–412

Robb EL, Winkelmolen L, Visanji N, Brotchie J, Stuart JA (2008b) Dietary resveratrol administration increases MnSOD expression and activity in mouse brain. Biochem Biophys Res Commun 372:254–259

Robb EL, Stuart JA (2011) Resveratrol interacts with estrogen receptor-β to inhibit cell replicative growth and enhance stress resistance by upregulating mitochondrial superoxide dismutase. Free Radic Biol Med 50(7):821–831

Rogers NH, Witczak CA, Hirshman MF, Goodyear LJ, Greenberg AS (2009) Estradiol stimulates Akt, AMP-activated protein kinase (AMPK) and TBC1D1/4, but not glucose uptake in rat soleus. Biochem Biophys Res Commun 382:646–650

Sarsour EH, Kalen AL, Xiao Z, Veenstra TD, Chaudhuri L, Venkataraman S, Reigan P, Buettner GR, Goswami PC (2012) Manganese superoxide dismutase regulates a metabolic switch during the mammalian cell cycle. Cancer Res 72(15):3807–3816

Schleipen B, Hertrampf T, Fritzemeier KH, Kluxen FM, Lorenz A, Molzberger A, Velders M, Diel P (2011) ERβ-specific agonists and genistein inhibit proliferation and induce apoptosis in the large and small intestine. Carcinogenesis 32(11):1675–1683

Schini-Kerth VB, Auger C, Etienne-Selloum N, Chataigneau T (2010) Polyphenol-induced endo-thelium-dependent relaxations role of NO and EDHF. Adv Pharmacol 60:133–175

Seifert EL, Caron AZ, Morin K, Coulombe J, He XH, Jardine K, Dewar-Darch D, Boekelheide K, Harper ME, McBurney MW (2012) SirT1 catalytic activity is required for male fertility and metabolic homeostasis in mice. FASEB J 26:555–566

Sequeira J, Boily G, Bazinet S, Saliba S, He X, Jardine K, Kennedy C, Staines W, Rousseaux C, Mueller R, McBurney MW (2008) Sirt1-null mice develop an autoimmune-like condition. Exp Cell Res 314:3069–3074

Shin SM, Cho IJ, Kim SG (2009) Resveratrol protects mitochondria against oxidative stress through AMP-activated protein kinase-mediated glycogen synthase kinase-3beta inhibition downstream of poly(ADP-ribose)polymerase-LKB1 pathway. Mol Pharmacol 76:884–895

Simão F, Matté A, Matté C, Soares FM, Wyse AT, Netto CA, Salbego CG (2011) Resveratrol prevents oxidative stress and inhibition of Na(+)K(+)-ATPase activity induced by transient global cerebral ischemia in rats. J Nutr Biochem 22(10):921–928

Simpkins JW, Yi KD, Yang SH, Dykens JA (2010) Mitochondrial mechanisms of estrogen neuro-protection. Biochim Biophys Acta 1800:1113–1120

Simpkins JW, Singh M, Brock C, Etgen AM (2012) Neuroprotection and estrogen receptors. Neuroendocrinology 96:119–130

Sivritas D, Becher MU, Ebrahimian T, Arfa O, Rapp S, Bohner A, Mueller CF, Umemura T, Wassmann S, Nickenig G, Wassmann K (2011) Antiproliferative effect of estrogen in vascular smooth muscle cells is mediated by Kruppel-like factor-4 and manganese superoxide dis-mutase. Basic Res Cardiol 106(4):563–575

Stirone C, Duckles SP, Krause DN, Procaccio V (2005) Estrogen increases mitochondrial effi-ciency and reduces oxidative stress in cerebral blood vessels. Mol Pharmacol 68:959–965

Strehlow K, Rotter S, Wassmann S, Adam O, Grohé C, Laufs K, Böhm M, Nickenig G (2003) Modulation of antioxidant enzyme expression and function by estrogen. Circ Res 93(2):170–177

Sugiyama N, Barros RP, Warner M, Gustafsson JA (2010) ERbeta: recent understanding of estro-gen signaling. Trends Endocrinol Metab 21:545–552

Al Sweidi S, Sánchez MG, Bourque M, Morissette M, Dluzen D, Di Paolo T (2012) Oestrogen receptors and signalling pathways: implications for neuroprotective effects of sex steroids in Parkinson's disease. J Neuroendocrinol 24:48–61

Timmers S, Konings E, Bilet L, Houtkooper RH, van de Weijer T, Goossens GH, Hoeks J, van der Krieken S, Ryu D, Kersten S, Moonen-Kornips E, Hesselink MK, Kunz I, Schrauwen-Hinderling VB, Blaak EE, Auwerx J, Schrauwen P (2011) Calorie restriction-like effects of 30 days of resveratrol supplementation on energy metabolism and metabolic profile in obese humans. Cell Metab 14:612–622

Tissenbaum HA, Guarente L (2001) Increased dosage of a sir-2 gene extends lifespan in Caenorhabditis elegans. Nature 410:227–230

Ungvari Z, Labinskyy N, Mukhopadhyay P, Pinto JT, Bagi Z, Ballabh P, Zhang C, Pacher P, Csiszar A (2009) Resveratrol attenuates mitochondrial oxidative stress in coronary arterial endothelial cells. Am J Physiol Heart Circ Physiol 297(5):H1876–H1881

Valenzano DR, Terzibasi E, Genade T, Cattaneo A, Domenici L, Cellerino A (2006) Resveratrol prolongs lifespan and retards the onset of age-related markers in a short-lived vertebrate. Curr Biol 16:296–300

Vingtdeux V, Giliberto L, Zhao H, Chandakkar P, Wu Q, Simon JE, Janle EM, Lobo J, Ferruzzi MG, Davies P, Marambaud P (2010) AMP-activated protein kinase signaling activation by resveratrol modulates amyloid-beta peptide metabolism. J Biol Chem 285:9100–9113

Wang L, Andersson S, Warner M, Gustafsson JA (2001) Morphological abnormalities in the brains of estrogen receptor beta knockout mice. Proc Natl Acad Sci USA 98:2792–2796

Wang M, Crisostomo PR, Markel T, Wang Y, Lillemoe KD, Meldrum DR (2008) Estrogen receptor beta mediates acute myocardial protection following ischemia. Surgery 144:233–238

Wang M, Wang Y, Weil B, Abarbanell A, Herrmann J, Tan J, Kelly M, Meldrum DR (2009) Estrogen receptor beta mediates increased activation of PI3K/Akt signaling and improved myocardial function in female hearts following acute ischemia. Am J Physiol Regul Integr Comp Physiol 296(4):R972–R978

Westerheide SD, Anckar J, Stevens SM Jr, Sistonen L, Morimoto RI (2009) Stress-inducible regulation of heat shock factor 1 by the deacetylase SIRT1. Science 323:1063–1106

Whyte L, Huang YY, Torres K, Mehta RG (2007) Molecular mechanisms of resveratrol action in lung cancer cells using dual protein and microarray analyses. Cancer Res 67:12007–12017

Wilson BJ, Tremblay AM, Deblois G, Sylvain-Drolet G, Giguère V (2010) An acetylation switch modulates the transcriptional activity of estrogen-related receptor alpha. Mol Endocrinol 24:1349–1358

Wood JG, Rogina B, Lavu S, Howitz K, Helfand SL, Tatar M, Sinclair DA (2004) Sirtuin activators mimic caloric restriction and delay ageing in metazoans. Nature 430:686–689

Yao J, Hamilton RT, Cadenas E, Brinton RD (2010a) Decline in mitochondrial bioenergetics and shift to ketogenic profile in brain during reproductive senescence. Biochim Biophys Acta 1800:1121–1126

Yao J, Chen S, Cadenas E, Brinton RD (2010b) Estrogen protection against mitochondrial toxin-induced cell death in hippocampal neurons: antagonism by progesterone. Brain Res 1379:2–10

Yao Y, Li H, Gu Y, Davidson NE, Zhou Q (2010c) Inhibition of SIRT1 deacetylase suppresses estrogen receptor signaling. Carcinogenesis 31:382–387

Yao J, Irwin R, Chen S, Hamilton R, Cadenas E, Brinton RD (2012) Ovarian hormone loss induces bioenergetic deficits and mitochondrial β-amyloid. Neurobiol Aging 33(8):1507–1521

Yepuru M, Eswaraka J, Kearbey JD, Barrett CM, Raghow S, Veverka KA, Miller DD, Dalton JT, Narayanan R (2010) Estrogen receptor-{beta}-selective ligands alleviate high-fat diet- and ovariectomy-induced obesity in mice. J Biol Chem 285:31292–31303

Yoshino J, Conte C, Fontana L, Mittendorfer B, Imai S, Schechtman KB, Gu C, Kunz I, Rossi Fanelli F, Patterson BW, Klein S (2012) Resveratrol supplementation does not improve metabolic function in nonobese women with normal glucose tolerance. Cell Metab 16:658–664

Yuan P, Liang K, Ma B, Zheng N, Nussinov R, Huang J (2011) Multiple-targeting and conformational selection in the estrogen receptor: computation and experiment. Chem Biol Drug Des 78(1):137–149

Zang M, Xu S, Maitland-Toolan KA, Zuccollo A, Hou X, Jiang B, Wierzbicki M, Verbeuren TJ, Cohen RA (2006) Polyphenols stimulate AMP-activated protein kinase, lower lipids, and inhibit accelerated atherosclerosis in diabetic LDL receptor-deficient mice. Diabetes 55:2180–2191

Zheng J, Ramirez VD (1999) Rapid inhibition of rat brain mitochondrial proton F0F1-ATPase activity by estrogens: comparison with Na+, K+-ATPase of porcine cortex. Eur J Pharmacol 368:95–102

Zheng J, Ramirez VD (2000) Inhibition of mitochondrial proton F0F1-ATPase/ATP synthase by polyphenolic phytochemicals. Br J Pharmacol 130:1115–1123

Chapter 4
Bioavailability of Resveratrol, Pterostilbene, and Piceid

4.1 Introduction

There is generally good agreement between the in vitro effects of resveratrol and derivatives observed in cell culture experiments using low micromolar concentrations and those observed in vivo using either dietary supplementation or injection. This is somewhat surprising given the consistent observation of very low resveratrol bioavailability in vivo (see below). Resveratrol is rapidly and extensively metabolized in vivo, such that concentrations measured in human or rodent plasma following oral administration typically range from high nanomolar to just a few micromolar. In this chapter we present and review the available data on resveratrol bioavailability in humans and rodents taking these polyphenols as dietary supplements or via alternative administration routes. We attempt to reconcile the relationships between concentrations and activities observed in vitro and in vivo. The more limited information regarding bioavailability of resveratrol derivatives is reviewed. Further, the relatively recent literature on alternative delivery methods to boost bioavailability is explored.

4.2 Resveratrol Metabolism In Vivo

Resveratrol undergoes extensive chemical modification in the intestinal tract and is rapidly metabolized. In humans, plasma levels of resveratrol following a single 25 mg oral dose peak at an average concentration of 2 μM, but the vast majority of this is metabolized derivatives of unknown biological activity (Walle et al. 2004). In rodents oral intake of resveratrol in the hundreds of milligrams range also yields only low nanomolar plasma levels of unmetabolized resveratrol (Teng et al. 2012; Marier et al. 2002). Tissue levels of resveratrol in rodents following a high oral dose are also in the nanomolar range, with the highest concentrations being observed in liver and kidney tissue (Juan et al. 2010). Resveratrol is capable of crossing the

J.A. Stuart and E.L. Robb, *Bioactive Polyphenols from Wine Grapes*,
SpringerBriefs in Cell Biology, DOI 10.1007/978-1-4614-6968-1_4,
© The Author(s) 2013

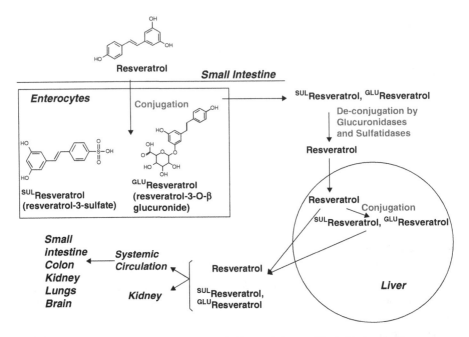

Fig. 4.1 Outline of resveratrol metabolism to sulfate and glucuronide derivatives in vivo

blood–brain barrier, and pure resveratrol has been measured in brain tissues of rats and mice given oral doses of resveratrol (Juan et al. 2010; Vitrac et al. 2003) (Fig. 4.1).

The limitations of resveratrol's very low bioavailability can be at least partially overcome by increasing dose. Whereas a single 25 mg dose of resveratrol, corresponding to moderate to high red wine consumption, resulted in marginal levels of plasma resveratrol in human subjects, a single 5 g dose increased this to up to 2.4 μM (Boocock et al. 2007). In a second study by the same group, Brown et al. (2010) showed that daily consumption of the same 5 g dose of resveratrol for several weeks further increased peak plasma concentrations to almost 5 μM. In a separate long-term study using lower daily doses of resveratrol, Almeida et al. (2009) also showed that repeated dosing can increase the plasma half-life of resveratrol by more than twofold. Thus, the possibility exists that chronic dietary intake of resveratrol in several gram per day quantities could boost plasma resveratrol to levels in the low micromolar range that can elicit some effects in vitro. In considering these data, it should be noted also that in virtually all in vitro experiments resveratrol concentrations are typically 5–50 μM and the duration of treatment is almost always less than 1 week. In comparison, in vivo studies may not achieve equally high concentrations in plasma, but the duration of exposure is often much longer. How this impacts cellular interactions of resveratrol is not well enough understood, but presumably it would promote cellular uptake. In rabbits, rats and mice tissue levels of resveratrol following oral delivery closely parallel plasma levels (Asensi et al. 2002), suggesting that higher daily doses of resveratrol will be communicated into the extravascular tissues and might reach levels high enough to be effective. That this does indeed occur is strongly suggested by the simple observation that dietary intake of resveratrol at high levels can elicit many of the same biological activities observed in vitro (e.g., Chap. 2, Tables 2.1 and 2.2). In this context it is important to note that no serious adverse side effects have been associated with these higher levels of intake in humans (Edwards et al. 2011).

4.3   Concentrations of Resveratrol In Vitro vs. In Vivo

Although the concentrations of resveratrol required to elicit its effects in vitro are often several fold higher than those measured in vivo following dietary supplementation or other delivery methods, several factors affecting free polyphenol concentrations must be considered in interpreting these apparent differences. Resveratrol and many other red wine polyphenols are highly lipophilic and bind to serum albumin and other proteins that are contained in the fetal calf serum used in virtually all culture media. Thus, the free concentrations of polyphenols to which cells are exposed will be much lower than the initial dose. Jannin et al. (2004) showed that the concentration of free resveratrol in a common cell culture medium containing 10 % fetal calf serum follows an exponential extinction, reaching 50 % of the initial concentration added by 2 h, and falling to close to zero by 24 h. In the HepG2 cell line, a commonly used model for human hepatocytes, resveratrol uptake is thought to occur by a combination of passive transport and simple diffusion. Resveratrol uptake by simple diffusion is nearly twofold lower when serum is included in the culture medium, again suggesting that serum protein binding affects resveratrol's cellular uptake (Delmas and Lin 2011). Thus, it is likely that an addition of 50 μM resveratrol in an experiment lasting 24 h might give an average concentration over the duration of the experiment in the low micromolar range. The presence of serum in culture media is thus a confounding factor that limits comparisons that must be considered when comparing in vitro and in vivo concentrations.

A second factor that influences resveratrol uptake in vitro and must be considered when making comparisons with in vivo experiments is the choice of vehicle solvent for resveratrol addition. Resveratrol is soluble in ethanol and dimethylsulfoxide (DMSO), and both are commonly used vehicles to deliver resveratrol in vitro. Ethanol, but not DMSO, can induce an expansion of lipid membranes that alters fluidity and this modifies resveratrol uptake. The concentration of resveratrol needed to observe its anti-proliferative effects in HepG2 cells is lower when ethanol is used as a solvent than with DMSO as a solvent (Delmas et al. 2000). Thus, specific properties of the culture medium including the means of introducing resveratrol (or other polyphenols) complicate the direct comparison of in vitro and in vivo resveratrol concentrations.

4.4   In Vivo Bioavailability of Resveratrol Metabolites and Derivatives

The low bioavailability of resveratrol in vivo is due largely to its rapid metabolism following ingestion, and the majority of the resveratrol detected in animal plasma and tissues exists as the sulfate and glucuronide metabolites (Wenzel and Somoza 2005). Wenzel et al. (2005) fed rats a 300 mg/kg body weight/day dose of resveratrol for 8 weeks. While plasma levels of free resveratrol were less than 0.3 μg/mL (~1.3 μM), the resveratrol-3-sulfate and glucuronide derivatives were detected at 0.37 ± 0.09 μg/mL (~1.6 μM) and 3.13 ± 0.88 μg/mL (~15 μM), respectively. Given that oral doses of resveratrol have been shown to be effective in inhibiting the growth and progression of cancer, one might hypothesize that resveratrol's more bioavailable metabolites contribute to this biological activity. However, in vitro experiments to date have used pure resveratrol almost exclusively, so only a few studies shed light on the possible biological activities of its most prominent metabolites. We investigated the in vitro effects of resveratrol-3-sulfate and resveratrol-4-glucuronide, and found that unlike resveratrol, these compounds had no effect on cell proliferation rates in a mouse myoblast cell line (C2C12) and a human lung fibroblast cell line (MRC5) at concentrations up to 100 μM (Robb and Stuart, unpublished). In corroboration with our observations in myoblasts and fibroblasts, the proliferative rate of the cancerous mammary cell line MC-7 was similarly unaffected by the 3′-sulfate and 4-sulfate resveratrol metabolites (Miksits et al. 2009). In contrast, however, sulfate metabolites were shown to have biological activity in cultured macrophage cells where, similar to what is observed with resveratrol, the 3′-sulfate, 3,4′-sulfate, and 3-sulfate metabolites inhibited the activity of inducible nitric oxide synthase (Hoshino et al. 2010). The resveratrol-3-sulfate and glucuronide metabolites, at concentrations of 10 μM, also exert effects on 3T3-L1 adipocytes in vitro (Lasa et al. 2012). Thus, resveratrol metabolites may explain some of the biological activities in vivo that have been associated with resveratrol supplementation. A complete evaluation of the biological activities of resveratrol metabolites in different experimental contexts is an important step toward a more complete and accurate understanding of resveratrol's in vivo effects.

Much of the existing literature surrounding red wine polyphenols has focused on isolated resveratrol, and the activities of related compounds or potential synergistic effects with the many other polyphenols found in red wine are a newer area of research. The potential activity of pterostilbene, a methylated derivative of resveratrol found in red wine, gained research interest recently after it was found to have anti-proliferative effects in cultured cells at lower concentrations than resveratrol (e.g., McCormack et al. 2011). In vitro, the concentration of pterostilbene required to reduce the proliferative growth rate of human colon cancer cell lines (HCT116, HT29, Caco-2) is approximately half that required for resveratrol (Nutakul et al. 2011). In addition to colon cancer cell lines, pterostilbene inhibits the proliferation of preadipocytes and breast cancer cell lines (Hsu et al. 2012; Lee et al. 2011; McCormack et al. 2011). In vivo, inclusion of pterostilbene in the diet of mice significantly decreased rates of pancreatic tumor growth (McCormack et al. 2012). In rats, oral administration of pterostilbene (10 mg/kg) resulted in higher serum concentrations (up to 100 ng/ml, or 0.4 μM) and plasma exposure, and lower clearance than resveratrol at similar doses (Lin et al. 2009). A second rat study using up to 2 weeks of 170 mg/kg/day pterostilbene versus 150 mg/kg/day resveratrol showed that pterostilbene is up to tenfold more bioavailable, reaching peak plasma concentrations of over 20 μM. Some researchers have suggested that pterostilbene's methyl groups might prevent the extensive metabolism observed with resveratrol in vivo, thus facilitating the higher circulating concentrations of pterostilbene (Wen and Walle 2006; Nutakul et al. 2011). In any case, these results certainly highlight pterostilbene as a potentially more attractive candidate for further development than resveratrol, particularly given its greater potency in some in vitro assays of biological activity. Another abundant wine polyphenol, piceatannol, also shows in vivo biological activities with promising health effects, such as the ability to reduce infarct size and neural deficits associated with middle cerebral artery occlusion in male rats (Wang et al. 2012). However, there is currently insufficient data to evaluate its bioavailability in vivo.

Several other viniferins have recently been identified for their potential therapeutic use. For example, ε-viniferin causes a greater inhibition of proliferation in isolated rat vascular smooth muscle cells than resveratrol at equivalent doses (Zghonda et al. 2011). In cultured neuronal cells (PC12) ε-viniferin is protective against the toxic effects of the β-amyloid peptide, and is thus thought to have neuroprotective properties (Richard et al. 2011). However, there is as yet no data available regarding the bioavailability of viniferins in vivo.

4.5 Resveratrol Delivery Strategies to Increase Bioavailability

Two important factors impinging upon the bioavailability and in vivo activities of resveratrol and other red wine polyphenols are the delivery method and vehicle used. This is particularly true when considering data generated using rodent models, where a wider breadth of delivery methods is available that may not be plausible in humans. In the majority of studies using rodent models resveratrol is incorporated into the animal's chow or drinking water, thus providing a chronic low dose exposure. In contrast, in humans the current studies have relied on supplements to achieve the desired resveratrol intake. This is an important distinction from the animal studies, as a supplement model more closely represents a bolus dose of resveratrol than chronic exposure.

The composition of the delivery matrix is also likely to impact the bioavailability of resveratrol and derivatives. In dietary resveratrol administration, significant effects on metabolism and aerobic exercise endurance are observed when resveratrol is added to diets with fat contents equal to or greater than 40 %, but not when resveratrol is added to a standard composition diet (Baur et al. 2006; Lagouge et al. 2006). Similarly, resveratrol given in a high-fat diet increases brain antioxidant enzyme activities by approximately twofold, while the same dose given in a standard mouse diet does not have a significant effect (Robb et al. 2008b). An often-overlooked factor in these studies is the fat source used to augment the animal diet. The use of oils high in monounsaturated and polyunsaturated fatty acids to increase the fat component of the diet can have substantial physiological effects that may alter the response to resveratrol (Lau et al. 2010). Alternatively, however, it may simply be that the high-fat diet directly affects the intestinal absorption, metabolism, and transport of resveratrol, which is a highly lipophilic compound. Resveratrol undergoes extensive chemical modification in the intestinal tract, and its inclusion in a high-fat diet may alter the degree to which it is modified (Kuhnle et al. 2000).

Attempts to promote resveratrol delivery and uptake using specific matrices have as yet yielded limited success (reviewed in Santos et al. 2011; Amri et al. 2012). To circumvent the negative consequences of a high-fat diet, we developed a silicon-based encapsulation formula as a means of reducing chemical modifications of resveratrol during its ingestion. However, at a matching resveratrol dose the silicon formulation failed to replicate the effects on tissue antioxidant levels and mitochondrial abundance observed in mice when resveratrol is given in combination with a high-fat diet (Robb et al., unpublished). Alternative methods of shielding resveratrol from intestinal modifications have also been reported including encapsulation in a nanoparticle delivery system, liposome encapsulation, and the synthesis of more bioavailable derivatives (see Teskac and Kristl 2010; Gokce et al. 2012; Coimbra et al. 2011), but their ability to increase resveratrol's bioavailability has not been thoroughly tested in vivo.

In light of these observations of an apparent requirement for delivery in a high-fat diet to elicit many of resveratrol's effects in vivo, it would seem possible that dietary strategies could be used to improve resveratrol's bioavailability in humans. However, to date, little research exploring this idea has been published. Vitaglione and colleagues (2005) observed that circulating levels of resveratrol following red wine intake were not influenced by the macromolecule composition of an accompanying meal in healthy human subjects. While this result does not support the argument that diet composition influences resveratrol uptake, any conclusions based on this study are limited by the fact that the plasma concentrations of resveratrol detected in this study were extremely low, making the data difficult to interpret. Understanding the impact of diet composition on resveratrol's effects is important for its potential to inform nutritional strategies and the development of new delivery methods with the goal of increasing bioavailability, and will be an area of ongoing research.

4.6 Conclusions

In summary, it is clear that, while the relatively low resveratrol levels present in red wines are less likely to strongly elicit the biological effects observed in in vitro studies, dietary supplementation with higher doses might provide an effective plasma concentration. The ongoing publication of data indicating that pterostilbene and other red wine polyphenols have similar activities in cultured cells strongly suggests that resveratrol is not unique but rather one of several grapevine compounds with beneficial effects. Since some of these other compounds have greater bioavailability in vivo than resveratrol, future research should focus on more fully characterizing them, particularly pterostilbene and piceid.

References

Almeida L, Vaz-da-Silva M, Falcao A, Soares E, Costa R, Loureiro AI, Fernandes-Lopes C, Rocha JF, Nunes T, Wright L, Soares-da-Silva P (2009) Pharmacokinetic and safety profile of trans-resveratrol in a rising multiple-dose study in healthy volunteers. Mol Nutr Food Res 53:S7–S15

Amri A, Chaumeil JC, Sfar S, Charrueau C (2012) Administration of resveratrol: what formulation solutions to bioavailability limitations? J Control Release 158:182–193

Asensi M, Medina I, Ortega A, Carretero J, Bano MC, Obrador E, Estrela JM (2002) Inhibition of cancer growth by resveratrol is related to its low bioavailability. Free Radic Biol Med 33:387–398

Baur JA, Pearson KJ, Price NL, Jamieson HA, Lerin C, Kalra A, Prabhu VV, Allard JS, Lopez-Lluch G, Lewis K, Pistell PJ, Poosala S, Becker KG, Boss O, Gwinn D, Wang M, Ramaswamy S, Fishbein KW, Spencer RG, Lakatta EG, Le Couteur D, Shaw RJ, Navas P, Puigserver P, Ingram DK, de Cabo R, Sinclair DA (2006) Resveratrol improves health and survival of mice on a high-calorie diet. Nature 444:337–342

Boocock DJ, Faust GE, Patel KR, Schinas AM, Brown VA, Ducharme MP, Booth TD, Crowell JA, Perloff M, Gescher AJ, Steward WP, Brenner DE (2007) Phase I dose escalation pharmacokinetic study in healthy volunteers of resveratrol, a potential cancer chemopreventive agent. Cancer Epidemiol Biomarkers Prev 16:1246–1252

Brown VA, Patel KR, Viskaduraki M, Crowell JA, Perloff M, Booth TD, Vasilinin G, Sen A, Schinas AM, Piccirilli G, Brown K, Steward WP, Gescher AJ, Brenner DE (2010) Repeat dose study of the cancer chemopreventive agent resveratrol in healthy volunteers: safety, pharmacokinetics, and effect on the insulin-like growth factor axis. Cancer Res 70:9003–9011

Coimbra M, Isacchi B, van Bloois L, Torano JS, Ket A, Wu X, Broere F, Metselaar JM, Rijcken CJ, Storm G, Bilia R, Schiffelers RM (2011) Improving solubility and chemical stability of natural compounds for medicinal use by incorporation into liposomes. Int J Pharm 416:433–442

Delmas D, Lin HY (2011) Role of membrane dynamics processes and exogenous molecules in cellular resveratrol uptake: consequences in bioavailability and activities. Mol Nutr Food Res 55:1142–1153

Delmas D, Jannin B, Cherkaoui Malki M, Latruffe N (2000) Inhibitory effect of resveratrol on the proliferation of human and rat hepatic derived cell lines. Oncol Rep 7:847–852

Edwards JA, Beck M, Riegger C, Bausch J (2011) Safety of resveratrol with examples for high purity, trans-resveratrol, resVida. Ann NY Acad Sci 1215:131–137

Gokce EH, Korkmaz E, Dellera E, Sandri G, Bonferoni MC, Ozer O (2012) Resveratrol-loaded solid lipid nanoparticles versus nanostructured lipid carriers: evaluation of antioxidant potential for dermal applications. Int J Nanomedicine 7:1841–1850

Hoshino J, Park EJ, Kondratyuk TP, Marler L, Pezzuto JM, van Breemen RB, Mo S, Li Y, Cushman M (2010) Selective synthesis and biological evaluation of sulfate-conjugated resveratrol metabolites. J Med Chem 53:5033–5043

Hsu CL, Lin YJ, Ho CT, Yen GC (2012) Inhibitory effects of garcinol and pterostilbene on cell proliferation and adipogenesis in 3T3-L1 cells. Food Funct 3:49–57

Jannin B, Menzel M, Berlot JP, Delmas D, Lancon A, Latruffe N (2004) Transport of resveratrol, a cancer chemopreventive agent, to cellular targets: plasmatic protein binding and cell uptake. Biochem Pharmacol 68:1113–1118

Juan ME, Maijo M, Planas JM (2010) Quantification of trans-resveratrol and its metabolites in rat plasma and tissues by HPLC. J Pharm Biomed Anal 51:391–398

Kuhnle G, Spencer JP, Chowrimootoo G, Schroeter H, Debnam ES, Srai SK, Rice-Evans C, Hahn U (2000) Resveratrol is absorbed in the small intestine as resveratrol glucuronide. Biochem Biophys Res Commun 272:212–217

Lagouge M, Argmann C, Gerhart-Hines Z, Meziane H, Lerin C, Daussin F, Messadeq N, Milne J, Lambert P, Elliott P, Geny B, Laakso M, Puigserver P, Auwerx J (2006) Resveratrol improves mitochondrial function and protects against metabolic disease by activating SIRT1 and PGC-1alpha. Cell 127:1109–1122

Lasa A, Churruca I, Eseberri I, Andres-Lacueva C, Portillo MP (2012) Delipidating effect of resveratrol metabolites in 3T3-L1 adipocytes. Mol Nutr Food Res 56:1559–1568

Lau BY, Fajardo VA, McMeekin L, Sacco SM, Ward WE, Roy BD, Peters SJ, Leblanc PJ (2010) Influence of high-fat diet from differential dietary sources on bone mineral density, bone strength, and bone fatty acid composition in rats. Appl Physiol Nutr Metab 35:598–606

Lee MF, Pan MH, Chiou YS, Cheng AC, Huang H (2011) Resveratrol modulates MED28 (Magicin/EG-1) expression and inhibits epidermal growth factor (EGF)-induced migration in MDA-MB-231 human breast cancer cells. J Agric Food Chem 59:11853–11861

Lin HS, Yue BD, Ho PC (2009) Determination of pterostilbene in rat plasma by a simple HPLC-UV method and its application in pre-clinical pharmacokinetic study. Biomed Chromatogr 23:1308–1315

Marier JF, Vachon P, Gritsas A, Zhang J, Moreau JP, Ducharme MP (2002) Metabolism and disposition of resveratrol in rats: extent of absorption, glucuronidation, and enterohepatic recirculation evidenced by a linked-rat model. J Pharmacol Exp Ther 302:369–373

McCormack D, Schneider J, McDonald D, McFadden D (2011) The antiproliferative effects of pterostilbene on breast cancer in vitro are via inhibition of constitutive and leptin-induced Janus kinase/signal transducer and activator of transcription activation. Am J Surg 202:541–544

McCormack DE, Mannal P, McDonald D, Tighe S, Hanson J, McFadden D (2012) Genomic analysis of pterostilbene predicts its antiproliferative effects against pancreatic cancer in vitro and in vivo. J Gastrointest Surg 16:1136–1143

Miksits M, Wlcek K, Svoboda M, Kunert O, Haslinger E, Thalhammer T, Szekeres T, Jager W (2009) Antitumor activity of resveratrol and its sulfated metabolites against human breast cancer cells. Planta Med 75:1227–1230

Nutakul W, Sobers HS, Qiu P, Dong P, Decker EA, McClements DJ, Xiao H (2011) Inhibitory effects of resveratrol and pterostilbene on human colon cancer cells: a side-by-side comparison. J Agric Food Chem 59:10964–10970

Richard T, Poupard P, Nassra M, Papastamoulis Y, Iglesias ML, Krisa S, Waffo-Teguo P, Merillon JM, Monti JP (2011) Protective effect of ε-viniferin on β-amyloid peptide aggregation investigated by electrospray ionization mass spectrometry. Bioorg Med Chem 19:3152–3155

Robb EL, Winkelmolen L, Visanji N, Brotchie J, Stuart JA (2008b) Dietary resveratrol administration increases MnSOD expression and activity in mouse brain. Biochem Biophys Res Commun 372:254–259

Santos AC, Veiga F, Ribeiro AJ (2011) New delivery systems to improve the bioavailability of resveratrol. Expert Opin Drug Deliv 8:973–990

Teng Z, Yuan C, Zhang F, Huan M, Cao W, Li K, Yang J, Cao D, Zhou S, Mei W (2012) Intestinal absorption and first-pass metabolism of polyphenol compounds in rat and their transport dynamics in Caco-2 cells. PLoS One 7:e29647

Teskac K, Kristl J (2010) The evidence for solid lipid nanoparticles mediated cell uptake of resveratrol. Int J Pharm 390:61–69

Vitaglione P, Sforza S, Galaverna G, Ghidini C, Caporaso N, Vescovi PP, Fogliano V, Marchelli R (2005) Bioavailability of trans-resveratrol from red wine in humans. Mol Nutr Food Res 49:495–504

Vitrac X, Desmouliere A, Brouillaud B, Krisa S, Deffieux G, Barthe N, Rosenbaum J, Merillon JM (2003) Distribution of [14C]-trans-resveratrol, a cancer chemopreventive polyphenol, in mouse tissues after oral administration. Life Sci 72:2219–2233

Walle T, Hsieh F, DeLegge MH, Oatis JE Jr, Walle UK (2004) High absorption but very low bioavailability of oral resveratrol in humans. Drug Metab Dispos 32:1377–1382

Wang X, Song R, Bian HN, Brunk UT, Zhao M, Zhao KS (2012) Polydatin, a natural polyphenol, protects arterial smooth muscle cells against mitochondrial dysfunction and lysosomal destabilization following hemorrhagic shock. Am J Physiol 302:R805–R814

Wen X, Walle T (2006) Methylated flavonoids have greatly improved intestinal absorption and metabolic stability. Drug Metab Dispos 34:1786–1792

Wenzel E, Somoza V (2005) Metabolism and bioavailability of trans-resveratrol. Mol Nutr Food Res 49:472–481

Wenzel E, Soldo T, Erbersdobler H, Somoza V (2005) Bioactivity and metabolism of trans-resveratrol orally administered to Wistar rats. Mol Nutr Food Res 49:482–494

Zghonda N, Yoshida S, Araki M, Kusunoki M, Mliki A, Ghorbel A, Miyazaki H (2011) Greater effectiveness of ε-viniferin in red wine than its monomer resveratrol for inhibiting vascular smooth muscle cell proliferation and migration. Biosci Biotechnol Biochem 75:1259–1267

Chapter 5
General Discussion and Future Considerations

Over the past 20 years, the level of scientific and public interest in grapevine polyphenols has increased exponentially. Over that period of time, continually more potential human health applications of these molecules have been suggested and experimental data supporting various uses has accumulated. The earliest indications of anticancer and cardioprotective effects of resveratrol were followed by a succession of reports detailing resveratrol's anti-inflammatory, antiobesity, antiaging, antidiabetic, and neuroprotective properties. The level of interest in resveratrol remains so high that over 5,000 articles on the topic are listed in PubMed at the time of writing.

Perhaps not surprisingly then, resveratrol has generated considerable controversy amongst researchers. One of the most controversial areas has been the mechanism(s) of action of this molecule in mammalian cells. Early reports of resveratrol as a phytoestrogen appear to have fallen out of favor. Later reports of resveratrol as a direct activator of SirT1 have also been cast into doubt due to the unreliability of the protein deacetylase assay used (Pacholec et al. 2010; Beher et al. 2009; Borra et al. 2005; Kaeberlein et al. 2005). Putative direct interactions of resveratrol with AMPK and phosphodiesterase published more recently occur at concentrations that cannot be reached in vivo, so these mechanisms too are uncertain. Thus, it is not entirely clear how to reconcile the very low bioavailability of resveratrol in vivo with the relatively high concentrations being used generally to show effects in vitro. Nonetheless many of the in vitro effects of resveratrol can be reproduced in vivo.

Many of resveratrol's effects both in vitro and in vivo are virtually identical to those associated with 17β-estradiol. Thus, in our view the circumstantial evidence strongly suggests that an estrogenic mechanism contributes to the effects of resveratrol in mammalian cells and tissues. While this does not disclude other possible mechanisms, it must be realized that in many instances no other explanation seems necessary, since resveratrol largely phenocopies the effects of 17β-estradiol or one of the specific ERalpha or ERbeta agonists in a variety of contexts, including interactions with the proposed mechanisms outlined above (Table 5.1).

While resveratrol's inhibitory effects on cell growth appear to be primarily via ERbeta, which typically mediates anti-proliferative signals (Nilsson et al. 2011), it

J.A. Stuart and E.L. Robb, *Bioactive Polyphenols from Wine Grapes*,
SpringerBriefs in Cell Biology, DOI 10.1007/978-1-4614-6968-1_5,
© The Author(s) 2013

Table 5.1 Protective effects of resveratrol and estrogen in various disease contexts

Disease	Protection by resveratrol	Protection by estrogens
Atherogenesis	√	√
Hypertension	√	√
Ischemic stroke	√	√
Neurodegeneration	√	√
Cancer	√	Tissue specific
Obesity	√	√
Type 2 Diabetes	√	√

is able to bind both ERs, so as with 17β-estradiol itself, whether cell growth is stimulated or inhibited will depend in part upon the relative levels of ERalpha versus ERbeta. Other effects of resveratrol in the cardiovascular system and brain appear to be mediated by both ERalpha and ERbeta, so the overall effect in any tissue or cell is likely to reflect in part these relative receptor levels.

While resveratrol has captured the vast majority of attention amongst grapevine polyphenols, more recently attention has shifted to some of the resveratrol derivatives that may actually be produced in higher amounts in grapevines. Pterostilbene shares resveratrol's anti-proliferative and anticancer properties (McCormick & McFadden 2012; Robb and Stuart, unpublished), and is effective at lower concentrations than resveratrol in the limited studies that have been done to date. Pterostilbene bioavailability in vivo is also significantly higher than resveratrol, making it a potentially more attractive molecule for development as a health promoting nutraceutical. On the other hand, levels of pterostilbene in stressed grapevine leaves and grapes are lower than those of other polyphenols. Strategies to increase the production in grapes of this molecule in particular are therefore warranted. In general, on the basis of data collected to date, pterostilbene is a particularly interesting target for future development that will require additional research.

Piceid's effects in mammalian cells and tissues are also relatively understudied, although again the limited (relative to resveratrol) available data suggests this molecule can elicit many of the same effects as resveratrol and pterostilbene, perhaps via the same ER signalling pathways. At this time, there appears to be no data on piceid's bioavailability in vivo, which is an important deficit that should be addressed. However, it is one of the most abundant resveratrol derivatives found in the leaves and grapes of stressed grapevine, and thus is also one of the more abundant polyphenols present in red wines. Certainly, more research on the potential health effects of this molecule would shed light on its contribution to the effects of combined grapevine polyphenols and the potential for its further development as a nutraceutical.

The recent evidence of piceid and pterostilbene's biological activities in mammalian cells indicate the importance of considering all of these grapevine polyphenols, rather than focusing exclusively on resveratrol. While there are very few data on the effects of viniferins on mammalian biology, the small amount of available data is consistent with the hypothesis that these resveratrol oligomers are capable of eliciting some of the same cellular effects as resveratrol. Some viniferins are highly abundant in grapevine tissues, particularly under stressful conditions, and indeed they are captured at high concentrations in red wines. Currently, research on the viniferins is restricted by their very limited commercial availability, but it is hoped that this will change as interest in these molecules continues to grow. Also, although we have focused on the resveratrol derivatives in this work, there are dozens of additional polyphenolic molecules found in grapevine tissues that have not yet been well characterized and which deserve further study.

In red wines produced from relatively unstressed grapes using conventional fermentation approaches, the levels of resveratrol and its derivatives are probably so low, once their rapid in vivo metabolism is accounted for, that they could elicit marginal biological effects at best. Therefore, realizing significant health benefits will not likely be possible with moderate wine consumption unless these levels are significantly augmented. Continual refinements to the grape harvest and fermentation processes aimed at maximizing the production and extraction of resveratrol-based polyphenols is thus certainly worthwhile. Stimulation of the endogenous stilbene biosynthetic pathway using hormone elicitors like methyljasmonate shows some promise, as do various means of inducing stress. UV stress is a strong inducer of stilbene synthesis that could be exploited, perhaps even after grape harvest. Continuing research in this area is warranted to produce "high polyphenol" grapes and/or wines with proven health benefits. Although more research is needed on how higher levels of these molecules affect the palatability of red wines, the results from initial experiments with resveratrol suggest they will indeed be acceptable. Therefore, their inclusion at higher concentrations in red wines could yield health benefits without unduly compromising the sensory experience.

In conclusion, despite the wealth of data that has accumulated over two decades of studying grapevine polyphenols, there is much work still to do before the chemistry and biology of these compounds are sufficiently understood that they can be fully exploited as health promoting nutraceuticals. Many opportunities still exist for studying these molecules, particularly the many derivatives of resveratrol that have as yet been the focus of far less research effort than resveratrol itself. The limited available data for piceid, pterostilbene and viniferins available at this time are supportive of the hypothesis that they are potentially as effective as resveratrol at eliciting beneficial effects on health. Thus, even as human clinical studies on resveratrol in a variety of disease contexts are only recently underway, it is becoming apparent that the scope of study should expand to include these other molecules.

One caveat to the application of resveratrol and its derivatives in humans is that the majority of in vivo work on humans and animal models has been carried out in males or post-reproductive females. Given the possibility explored in Chap. 3 that many of the biological effects of resveratrol are via its activity as an estrogen agonist, its use in females of reproductive age is complicated. Similarly, since one of the most reproducible effects of resveratrol and its derivative is anti-proliferatiion, the potential for interference with developmental processes exists. This has been demonstrated for other phytoestrogens with similar structure, like genistein. This must be better studied in grapevine polyphenols before they are used in contexts where interference with growth and developmental could be problematic.

References

Beher D, Wu J, Cumine S, Kim KW, Lu SC, Atangan L, Wang M (2009) Resveratrol is not a direct activator of SIRT1 enzyme activity. Chem Biol Drug Des 74:619–624

Borra MT, Smith BC, Denu JM (2005) Mechanism of human SIRT1 activation by resveratrol. J Biol Chem 280:17187–17195

Kaeberlein M, McDonagh T, Heltweg B, Hixon J, Westman EA, Caldwell SD, Napper A, Curtis R, DiStefano PS, Fields S, Bedalov A, Kennedy BK (2005) Substrate-specific activation of sirtuins by resveratrol. J Biol Chem 280:17038–17045

McCormack D, McFadden D (2012) Pterostilbene and cancer: current review. J Surg Res 173:e53–e61

Nilsson S, Koehler KF, Gustafsson JÅ (2011) Development of subtype-selective oestrogen receptor-based therapeutics. Nat Rev Drug Discov 10:778–792

Pacholec M, Bleasdale JE, Chrunyk B, Cunningham D, Flynn D, Garofalo RS, Griffith D, Griffor M, Loulakis P, Pabst B, Qiu X, Stockman B, Thanabal V, Varghese A, Ward J, Withka J, Ahn K (2010) SRT1720, SRT2183, SRT1460, and resveratrol are not direct activators of SIRT1. J Biol Chem 285:8340–8351